Ian Shaw

Is it Safe to Eat?

Enjoy Eating and
Minimize Food Risks

Original illustrations by
Margaret Tanner

 Springer

Professor Dr. Ian Shaw
College of Science
University of Canterbury
Private Bag 4800
Christchurch 8020, New Zealand
ian.shaw@canterbury.ac.nz

ISBN 978-3-642-05962-9   e-ISBN 978-3-540-27003-4
DOI 10.1007/b 11381

Bibliographic information published by Die Deutsche Bibliothek
Die Deutsche Bibliothek lists this publication in the Deutsche Nationalbibliographie;
detailed bibliographic data is available in the Internet at <http://dnb.ddb.de>.

This work is subject to copyright. All rights are reserved, whether the whole or part of the
material is concerned, specifically the rights of translation, reprinting, reuse of illustrations,
recitation, broadcasting, reproduction on microfilm or in any other way, and storage in data
banks. Duplication of this publication or parts thereof is permitted only under the provi-
sions of the German Copyright Law of September 9, 1965, in its current version, and permission for use must always be obtained from Springer-Verlag. Violations are liable to pros-
ecution under the German Copyright Law.

Springer-Verlag is a part of Springer Science+Business Media
springeronline.com

1. © Ian Shaw, Lyttelton, New Zealand
2. © Springer-Verlag Berlin Heidelberg 2010
Printed in Germany

The use of general descriptive names, registered names, trademarks, etc. in this publication
does not imply, even in the absence of a specific statement, that such names are exempt
from the relevant protective laws and regulations and therefore free for general use.

Production editor: Christiane Messerschmidt, Rheinau
Cover: design & production GmbH, Heidelberg

Printed on acid-free paper   02/3020 – 5 4 3 2 1 0

Preface

The title of this book is a tribute to my mentor and friend, Professor André McLean, whose inaugural professorial lecture at University College London in 1981 was entitled *Is it safe to live?* In his talk he ran through life's toxicological risks and sparked my love affair with risk. I worked with him for 6 years during which time I consumed his knowledge and enthusiasm about food from both a risk and gastronomic perspective. So it seemed appropriate to adapt his title to food and to use this as the title for my first book about food.

The book is a personal journey through food safety. I tried to cover all of the burning issues in the field, and related these to my personal experiences. I have been involved in many of the case examples, the news headlines that I have used to illustrate the media's viewpoint were cut from newspapers at the time – I can't resist cutting out articles just in case I want to refer to them in the future. I usually don't, but the dog-eared yellowed cuttings in my box file labelled *interesting news clippings* gave a useful perspective for this book and justified their occupancy of my book shelf for over 20 years – not to mention their 14,000 mile journey from England to New Zealand when we emigrated 5 years ago.

It is difficult to get everything right, and to cover every angle of a subject as emotive as food safety especially when it is intended to reflect a personal viewpoint. I hope that the facts are right even if my interpretation of them is questionable to some – I intended to be controversial in places. Viewpoints make people think. Food safety has become such an enormous issue that people who often know nothing about it, e.g. politicians, express their opinion oh too often – so I thought that since I have worked in the field for 20 years that I too deserved a say!

The book is intended to give you an insight into the world of food safety and to help you look at the evidence about the effects of chemical and biological contaminants of food and make up your own mind about risk. I hope that it infuriates you in places, but that when you turn the last page that you feel you have a better understanding of one of the most important facets of our lives – FOOD.

By the way, I do answer the question *is it safe to eat?* but not until the last page of the book ... but don't take my word for it!

July 2004   Ian Shaw

Contents

Acknowledgements

I thank David Zehms for reading the manuscript with an intelligent non-scientist's mind and pointing out the numerous unintelligible sections, preparing the subject index (which is the key to a good book), and for his constant support and smiles during my incarceration in my study while writing the book.

Margaret Tanner added her characteristic undying efficiency to the preparation of the diagrams and to obtaining permission to reproduce the illustrations.

Many friends and colleagues have helped with diagrams and data – I particularly thank, Dr Andrew Hudson, Dr Ellen Podovinsky, Barbara Thomson and Gwyneth Carey-Smith.

I don't thank our neighbour for building a new house and causing considerable angst while I was trying to write. Writing with the hum of concrete pumping and nails being hammered into wood is an interesting intellectual challenge!

And finally, I thank New Zealand for being the most beautiful country in the world, and giving me an inspirational view from my study that makes me feel happy to be alive.

July 2004   Ian Shaw

To David

1
Food Safety Through the Ages

1   Food Safety Through the Ages

In prehistoric times cavemen's top priorities were getting food to eat and surviving. They did not have the intellect or the inclination to think whether the food was safe, and even if they had understood food safety they still would have been more concerned about eating to survive. Life itself was such a risky business that food associated problems paled into insignificance. Just going out to catch food was a significant life risk. Tracking down a huge beast, bringing it down and killing it posed an enormous threat to their personal safety. Imagine going into a field with a bull, wearing nothing but a loin cloth and carrying a piece of sharpened flint bound to a stick – the thought might be quite amusing, but when you are expected to kill the bull for lunch, I suspect that an air of seriousness – if not terror – might overcome you. This was everyday life for the average cave family between 70,000 and 200,000 years ago. It is likely that even faced with these unbelievable risks of living that if eating a particular food gave Neanderthal Man a stomach ache, or made him vomit that he would avoid that food in the future. We can't know this, but studies in animals have shown avoidance of foods containing toxins or organisms that might make them ill. This is learned behaviour and is very important for survival. If a caveman became ill he was very likely to die due to his inability to catch food, or because he could not run away from some marauding carnivorous animal or warring fellow human. So there was a significant survival advantage of being fit and healthy. Perhaps this was the birth of food safety, avoiding foods that resulted in illness because of their impact upon the individual's survival. But who knows? This is pure speculation.

The risks of prehistoric eating were two fold, being killed or injured catching animals to eat, or being harmed by toxic chemicals or disease causing microbes (pathogens) in food. Very little is known about all of this, but again a bit of speculation might not be a bad thing. Many of the plants of the time would have contained natural toxins to protect them from attack by insects and other herbivores – this is an important plant survival mechanism. Primitive man must have experimented to find out which vegetables and leaves were safe to eat. No doubt some diners were killed by the plants that they ate. As now there must have been plants around that were so toxic that just a couple of mouthfuls would be fatal (e.g. the Death Cap fungus (Amanita phalloides) of today). Clearly the people of the time must have learned to avoid these in order to survive. At the same time evolution was in progress and over the millennia between early man (Homo sapiens – thinking man) and Neanderthal Man it is possible that human metabolism evolved to allow toxic chemicals in food to be broken down and made safe. This was a key factor in human survival because it allowed people to eat a broader range of foods without succumbing to their toxicities. We have enzymes in our livers that break down toxic chemicals – one is called Cytochrome P450 – these enzymes are now considered of paramount importance by pharmacologists because they break down drugs and medicines as part of the process of elimination of these foreign chemicals from the body. But they evolved long before medicines were around. It is very likely that they evolved to detoxify food components as a means of protecting the consumer. Simple foods that we accept without thinking – such as cabbage – contain potentially highly toxic chemicals (e.g. glycoalkaloids) that are made safe by our liver's protective enzymes. Indeed the way that we digest and absorb food channels it straight to the liver to make certain that it is able to get rid of toxins before the absorbed nutrients are released to the rest of the body via the blood stream.

When we eat food its structure is broken down by mixing it with saliva and chewing in the mouth. The saliva also contains enzymes (amylases) that break down starch. Try chewing a small piece of bread for a few minutes, it will become sweeter as the sugars (the building blocks of starch) are released from the starch in the bread. The mush is then swallowed and goes via the gullet (oesophagus) to the stomach where it is exposed to a very strong acid (hydrochloric acid at pH 1) plus enzymes that break down proteins (peptidases). The peptidases need acid conditions to work, but the strong acidity also kills bacteria and viruses and so protects us from some of the harmful living components of our food. Next the food is squirted into the duodenum where it is mixed with alkaline (pH 10) bile from the gall bladder. The pH change allows the next group of enzymes, the lipases – fat digesting enzymes – to get to work on what's left of the food. As the food bolus moves from the duodenum into the small intestine (ileum) it is almost completely broken down into its component parts. Fats are broken down to fatty acids and triglycerides, proteins are broken down to amino acids, complex sugars (e.g. starch) are broken down to their simple sugar building blocks (e.g. glucose). The only thing left intact is cellulose (fibre from plants) which helps the gut to get a grip on the food as it passes through. The food's movement along the small intestine takes a long time – it is about 7 metres long; food takes about 20 hours to pass along its entire length – during its slow movement along the windy tubular ileum, absorption of nutrients (sugars, simple fats and fatty acids, amino acids, vitamins, etc.) takes place. The absorbed nutrients are transported by a special blood system (the hepatic portal system) to the liver for detoxification and metabolism to release energy to fuel the body. What's left of the food, mainly cellulose and bacteria picked up from the intestine – bacteria are important occupants of the gut, they assist metabolism – moves slowly into the large intestine (colon) where water and minerals (e.g. calcium) are absorbed. The waste is then stored in the rectum before the urge to go to the toilet is great enough to result in its expulsion into the modern day toilet, or behind a prehistoric bush (Fig. 1-1).

This digestion process evolved long before *Homo sapiens*. Rats, mice, dogs, possums, and all other mammals have a very similar process. The only difference relates to adaptations

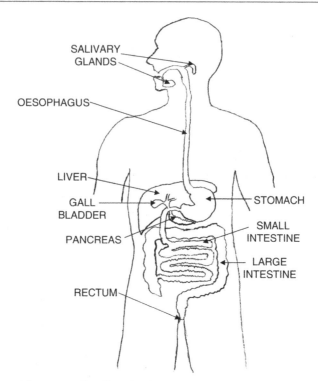

Fig. 1-1. The mammalian digestive system

to animals' differing diets. However different species do have different abilities to detoxify toxic components of food. Rabbits, for example, can eat small amounts of the deadly Death Cap fungus without coming to grief, whereas humans are very susceptible to its liver toxin (phalloidin – 200 mg will kill a human). This differential toxicity put cave men at grave danger, because they probably watched what animals ate before eating it themselves. This was a form of prehistoric animal testing. Life was trial and error in those times.

Eating animals is far less risky once you have killed them. This is because the animals' detoxifying systems have dealt with the poisons from their vegetable diets. Their flesh is therefore relatively free of toxic chemicals. Eating animals is a good way to get the nutrients from plants without suffering the potentially toxic outcomes of the plants. This is not always the case, for example polar bear liver is very rich in vitamin A due to their consumption of vitamin A-rich seals, who in turn consume vitamin A-rich fish. Concentration of the vitamin A in the livers up the food chain leads to toxic levels in polar bears. Cave men would have had similar problems with the animals that they ate. No doubt many died before their siblings and friends realised what had killed them, then avoided that food.

It probably did not take primitive men long to find out that when they caught their meat it could not be kept long before it did not taste good and made them not feel well. The concept of preserving food was evolved to reduce these problems. How this happened and when it happened is impossible to determine. One thing is certain, preserving did happen and played an important part in the survival and development of humans. Temperate countries where cold, often frozen, winters followed warm summers provided the right conditions for refrigeration preservation. Animals were probably more prevalent in the summer months than in winter. Prehistoric men might have noticed that carcasses lasted longer, and still tasted good and didn't give them prehistoric food poisoning in the winter months. It was only a small step for them to begin to stockpile their autumn catch in the winter snow to keep them and their families in food for the long winter months; this is speculation, but is still a possibility. In hot climates drying possibly evolved as a means of preserving fish and meat. Near to the sea salting developed. These three preservation methods are the same basic processes that we use today. Since bacteria evolved long before people appeared on earth, it is very likely indeed that the same basic bacterial food pathogens were the cause of prehistoric food poisoning as are the targets of our modern day ultra-clean food industry.

Therefore even though prehistoric people probably did not think about food in the same terms that we do today, they probably realised that they had to do certain things so as not to feel ill after eating. In this respect nothing has changed over the last 200,000 years since the first men walked the Earth.

Between 70,000 and 30,000 years ago, man changed very much indeed. They evolved socially from naked, or loin cloth-wearing, people to clothed (made from the skins of their food animals), advanced tool making, social people. But from the point of view of food safety the most important change was the introduction of cooking. No one knows when or why people first began to cook. They might have used fire to remove the coarse hairs from the skins of their food animals to make them more palatable, and then noticed the nice roasted flavour of the flesh; who knows? This very primitive throw-it-in-the-fire approach to cooking evolved into spit roasting and boiling both meat and vegetables. These changes are evidenced by the wonderful evolution of cooking pots. Go to any museum and you will see examples of beautiful clay cooking pots. Cooking, of course, made food much safer from a microbiological point of view. No doubt as cooking was introduced food poisoning incidence declined. Whether the cooks of the time noted this is impossible to know. I suspect that they were more concerned with the greater tenderness and good flavours of their cooked meats and vegetables, but the fact that food safety was linked to other benefits meant that it was selected in.

By 8,000 years ago, Neolithic man was growing crops and keeping food animals captive; farming had evolved. His wife was cooking in clay pots and baking bread-like concoctions from the primitive wheat that they grew. They salted and cooled their food to keep it for leaner times, and had a very much better idea about what was good to eat, and what made them feel ill. There are still people alive in the more remote parts of the World who live a Neolithic life-style. Not because they choose to, but because they are so remote from the modern world that they have not evolved the social behaviour, tools, and life-styles that we so-called developed people have. It is therefore possible for us to get a good idea of the diseases that these people suffer and extrapolate this back to their ancestors of 8,000 years ago.

Let's jump 6,000 years to the time of Christ. Religion was an important tool of food safety. Whether you are religious or not is irrelevant, the impact of religion was great in a food safety context. The Old Testament includes some pretty good food safety rules. Don't eat pork, and don't mix milk with meat. Pork carries a nasty parasite – Trichinella which causes trichinosis in humans; if you mix meat and milk in a nice warm environment the milk provides a perfect culture medium for meat bacteria resulting in a potentially lethal microbiological cocktail. This is all good Biblical common sense. Whether the Old Testament instituted these rules as the first attempts at food safety legislation is questionable, but it is possible that they arose from the knowledge of the time surrounding the effects that might result from these culinary activities. Perhaps they thought that God was punishing them for eating these foods by making them ill. It doesn't matter what their reasons were, the advice was sensible and probably reduced food-related illness. Other religions have similar taboo foods and food combinations. Some have no modern day food safety explanation – they might have then – but others remain sensible to this day. The Koran, the divine word of God revealed by the Prophet Mohammed, divides food into two basic categories, halal or permitted, and haram or forbidden.

The Koran specifically forbids 6 food categories: blood, carrion, pork, intoxicating beverages prepared from grapes, intoxicating drugs, and food previously dedicated to idols. It is interesting that the Koran, like the Old Testament bans pork, perhaps trichinosis was a greater problem in years gone by than it is today. Eliminating carrion (dead putrefying flesh) from the diet makes good food safety sense, blood might also carry bacteriological and viral contamination so reducing its intake is a good idea too. Banning alcohol might not be so good from the point of view of food safety. Water is a source of many pathogenic organisms, this was particularly so in times gone by, drinking alcoholic beverages is a good way to consume sterile water. Monks began brewing for exactly this reason – water was undrinkable, mead (fermented honey) and beer were safe alternatives that had rather a nice effect that I suspect the Monks quite enjoyed!

Hindu dietary codes are surprisingly similar to those of the Old Testament and the Koran. Caraka Samheta, a Hindu physician who lived 3,500 years ago, listed unacceptable foods. Most are unidentifiable today, but some of his ideas make very sound food safety sense. Don't drink rainy season river water – this is a good idea because it is more likely to be polluted with dangerous bacteria. Don't eat the Black Gram Bean – it is possible that this contains toxic principles (lectins) found in many beans. He also listed many other plants and vegetable that are not known today, perhaps they were toxic too. The most notable Hindu dietary taboo is beef; there is no known reason for beef being excluded on food safety grounds then, but now people might support this Hindu decree since the advent of BSE and Escherichia coli 0157… but more of these later.

The Romans invaded Britain about 2000 years ago. They arrived to find no political structure to the country and a multitude of kingdoms continually fighting each other. They introduced order and infrastructure to the country. By this time food preparation had evolved significantly. Food was cooked, there were complex recipes, breads, pastries, rich sauces, stews, herbs and spices. Some of the food of the day is reflected in modern day dishes. The Romans took their own cuisine to Britain, a cuisine that had a significant consideration for food safety, Fig. 1-2.

The Romans used many highly flavoured herbs and spices, it seems that they liked their food to have a highly complex and strong taste. One of the reasons for this might involve, but not overcome, a food safety issue. The Romans used lead to line many of their cooking and storage vessels. A lot of their food was quite acid and therefore dissolved the lead. It is clear from studies on Roman bones that they had a high body burden of lead, indeed many of them must have been suffering from chronic lead poisoning. One of the symptoms of lead poisoning is altered taste, often with a metallic taste in the mouth. Perhaps they tried to disguise the metallic taste with strong herbs and spices, or perhaps their sense of taste was so poor due to lead poisoning that the only way that they could taste anything was to make it incredibly highly flavoured. Their desire for highly spiced food, for whatever reason, had a good knock-on effect. Many herbs and spices contain antibacterial chemicals (e.g. thyme contains the antiseptic thymol – often used in modern day dentist's antiseptic mouthwashes), and so their inclusion at high concentrations in Roman food probably reduced pathogen levels.

Preserving food was common in Roman times. The use honey to raise sugar content above that in which bacteria will grow was introduced (this is why we use sugar to preserve fruits

Fig. 1-2. A roman banquet

as jam) – interestingly, honey also contains antibacterial agents
that would have helped preserve the food. Salting was common
and smoking food was also introduced. Smoking is an important preservation technique it infuses preserving chemicals into
foods – usually fatty foods because the smoke preservatives are
soluble in fat. The Romans smoked meat, fish and cheeses. Food
was elaborate in the rich Roman's household. They would enter-
tain regularly and impress their guests with amazing culinary
creations. The Emperor Vitellius describes some amazing (and
to my mind disgusting) food that he dedicated to the Goddess
Minerva; pike liver, pheasants' and peacocks' brains, flamingo
tongues, and lamprey roe. No expense was spared to impress. I

11

imagine that many of his guests were ill the day after his banquets from sheer overindulgence rather than food poisoning. During this time the average native Briton was just about surviving on gruel made from wheat or oats boiled in water, flat breads and the occasional taste of meat.

Despite early people's efforts, whether intentional or not, to make food safe by reducing the number of bacteria they ate, there is no doubt that they consumed millions more bacteria than we do today. Food is so clean now that we are not exposed to many of the bacteria that our forefathers inadvertently took in with their food. Until quite recently the cleanliness of our diet was thought to be good – progress. We are now finding that this is not so. A few bacteria in our food are probably a very good thing. Our immune system keeps a watching brief on alien organisms that invade our bodies. It produces antibodies that help to neutralise and eliminate these unwanted creatures. The process of neutralising involved making antibodies against the target organism – these are complex protein molecules that recognise a feature (often another protein protruding from the organisms cell surface) of the foreign organism. The protein binds to the organism and signals to the body's excretory defence systems to get rid of the tagged creature. Interestingly some bacteria share some of their cell surface markers. The immune system has an in-built memory which means that once challenged it remembers the cell surface proteins of the challenger and is able to quickly produce antibodies if the body is invaded by the same organism at a later date – sometimes many years hence. This is the principle by which vaccines work. We challenge the body with a vaccine (e.g. TB vaccine) which comprises bits and pieces of the virus or bacterium (in the case of TB vaccine – bacterial cell walls) that we are trying to protect the body from. Antibodies are made, and memorised; if the body is challenged by the real organism at some time in the future, the immunological memory is switched on and antibodies very quickly made to neutralise the pathogen.

A pathogenic bacterium might have some markers in common with a harmless food contaminant. So if challenged with the harmless contaminant the body's immune system will produce antibodies that will cross react with the cell surface markers of the pathogen. So by eating the harmless bacterium you make yourself immune to the harmful one. Continuing with the vaccines analogy; Edward Jenner (1749–1823), a British doctor who lived in Gloucestershire noted that farm workers who contracted cow pox were less likely to suffer from smallpox than others. This led him to "invent" vaccines. He deliberately infected people with cowpox (the first vaccine – named after Vaccinia the Latin for cowpox) and showed that this decreased the incidence of smallpox. The reason for his findings – he did not know this – was that the cow pox virus cross reacted with the much more dangerous small pox virus.

Good bugs were likely to be common components of food in days gone by and so people were probably better protected against food pathogens than we are today because of our sterile approach to food. I will return to this in Chapter 3.

The Romans basically began modern cookery. Nothing much has changed since then – except perhaps the recipes (for which we should be thankful – I shudder at the thought of Flamingo tongues on toast for supper!).

Let's move on a few years to Merry Olde England in Tudor times (1405–1603). Food and war were arguably the highlights of the Tudor lifestyle. In upper class households kitchens were enormous with a large staff to maintain a gargantuan output of flamboyant food to impress their guests, and perhaps begin the period of obesity that we are still in. At the other end of the spectrum, poor people were living in poverty with little food which was probably still cooked over an open fire outside with a primary regard for filling tummies rather than enjoyment.

The Tudors are known to have washed their hands, often with sweet smelling rose water, before and after eating. Someone serving up a pie would take great care not to touch the slice intended for their guest. This is the first reference that I could find to washing hands – whether the Tudors did this with food safety in mind is impossible to know. It might simply have been a way of reducing their body odour and so making sitting next to one another at the dining table a more pleasant experience. Who knows? Whatever the reason, washing their hands in rose water was a good food safety activity and might have reduced the spread of some of the food born diseases of the time.

There was little change in food, its preparation and serving between Tudor times and the beginning of the Victorian era (1836–1901); the main changes involved refinements in kitchen infrastructure – the open fires of Tudor cooks were replaced by ranges and oven that could be temperature controlled. Despite these changes, the food from kitchens between the 1480s and the early 2000s would be recognisable and palatable today – as you will see from the recipes below:

Sodde eggs *(from a Tudor recipe)*

Eggs were boiled and served with a mustard sauce which was made by grinding up mustard seed, pepper and salt, added to melted butter and sugar to taste.

Mustard sauce *(from Mrs Beeton's Cookery Book, 1861)*
1 tblspn mustard
1 dsrtspn flour
1 oz butter
1 gill boiling water
1 teaspoon vinegar

Eggs in mustard sauce *(from the Constance Spry Cookery Book, 1956)*
Hard boiled eggs are served in a sauce made from:
1/4 pint mayonnaise
juice of 1/2 lemon
salt & pepper
sugar
mustard
1 tsp chopped dill
2 tblspn lightly whipped cream

The Victorians played an enormous part in introducing public health systems into Britain. Many of these ideas were taken to distant parts of the world as the subjects of Queen Victoria subjects colonised the globe. There is no doubt that Victorian cooks knew that disease could be spread by unhygienic processes. They knew that washing their hands was a way to minimise this. Again there was a great difference between the poor and rich classes. The latter were still concerned with abating their hunger, food related disease would have been a secondary consideration – if they were aware of it – to filling their families' stomachs.

Beeton's Household Management – edited by Mrs Isabella Beeton, first published in 1861, is the font of all knowledge relating to the Victorian household. On "the advantages of cleanliness" she says, "Health and strength cannot be long continued unless the skin – all the skin – is washed frequently with a sponge or other means." Surprisingly, she makes no mention in the entire tome of hygiene in the kitchen. There is, however, an underlying theme of cleanliness. Under the duties of the Scullery-maid, it is made clear that part of her duty was to "wash and scour all these places (entrance halls, passages, stairs which lead to the kitchen) twice a week, with the tables, shelves. and cupboards." Similarly under the Housemaid's duties Mrs Beeton proclaims, "Cleanliness is next to Godliness". And the Dairy-maid instructed to "suspend cloths soaked in chloride of lime" across the room in sultry weather. This practice would not only have made the room smell fresh, but would have released antiseptic chlorine into the atmosphere. Mrs Beeton lived in an era when the "Germ Theory" (i.e. that microorganisms cause disease) was put forward by Louis Pasteur in France (1857) and picked up by Joseph Lister in Edinburgh who introduced antiseptics (1867). No doubt these breakthroughs were reported in the newspapers and raised people's awareness of infections.

With the Victorians' awareness of disease and methods of disease prevention, and that food might be a source of infection, the scene was set for food safety to assume greater importance by design rather than fortuitously. Despite this, it took until 1953 when the Irish government included food safety and food hygiene regulations in their revised Food Act. This is the point at which it became illegal to sell and prepare unsafe food. And this was the point at which the food industry cleaned up their act – perhaps too much (remember what I said about good bugs).

So far, all of our considerations about food safety have related to microbiological contamination – except the possibility that prehistoric people might eat toxic plants. The late 1950s were the turning point in this respect. A foray into the food journals of 1953 reveals much talk about chemicals used in farming. Most of the reference are to increasing yields using pesticides and fertilisers; this is hardly surprising since the world had just emerged from war and short food supply. Who cared, or even considered, what these wonder chemicals might do to consumers when they meant that food was plentiful and cheap? This was heaven after the austerity of the past decade; the war years were filled with recipes for swede pie and meat substitutes … and suddenly roast beef and piles of roast potatoes were back on the Sunday lunch table. Then came Rachel Carson's Silent Spring in 1961. It told a frightening tale of how these saviours of modern day agriculture would destroy the environment as we knew it. They were toxic. They would be absorbed by animals and plants and stay with them forever. They would kill wild birds and insects. What would they do to people?

"The question of chemical residues on the food we eat is a hotly debated issue" said Rachel Carson in Silent Spring – and it still is. It was in the early 1960s that concern over chemicals in food was born. Since then significant control measures have been introduced world-wide to control chemical residues in food. Maximum Residue Levels (MRLs) are now internationally accepted as a means of maintaining residues at a minimum based on the proper use of pesticides and other agrochemicals in farming. Food Acts around the world were revised in the decades between 1970 and 2000 to include chemical residues. In these mid years of the 1960s we were entering an era of food safety awareness, of legislative control measures, of public concern about the safety of the food they ate.

Public concern about food safety was heightened in the UK in 1988 when the then UK Minister of Health announced that poultry were contaminated with Salmonella and that this was linked to human disease associated with eating eggs.

Currie bows to ultimatum in egg affair – Mrs Edwina Currie has been forced to face MPs' questions today on her role in the Salmonella and eggs affair.

The Times, February 8, 1989

The UK egg industry collapsed and the press had a field day. The era of press hype around food safety issues had dawned. From that time forward few food safety issues, however small, escaped the pages of the daily newspapers. In fact some might even have been fuelled by the press – after all everyone was concerned about their food, and a good story sells newspapers! On the positive side, this heightened awareness and concern about food safety almost certainly led to increased government concern, and is very likely to have resulted in more regulations to ensure food safety. This effect was not peculiar to the UK, indeed most countries in the developed world became food safety aware in the 1980s.

It is important to remember that food safety issues affect everyone in the World, but still most people are not concerned about the safety of their food because their primary consideration is getting something to eat. Fifty percent of the world is still undernourished – this is a terrible state of affairs in a world that includes countries which overproduce food to an embarrassing extent. Surely we could re-distribute our food better.

In the late 1960s, there was an increased interest in food safety issues in the USA. This was the time of space travel, of enormous excitement about landing on distant planets and our own moon, and being the first nation to travel far into space. It was politically extremely important. This travel not only impressed Americans who became, quite rightly, very proud of their country's achievements. But more than this it showcased the USA's intellectual and scientific ability to the USSR – these were the high years of the Cold War. Star Wars was just around the corner. The USA realised that if anything went wrong with one of their manned space missions that they would lose face both at home and with their USSR opponents. In the context of space travel it was a very mundane issue that concerned them. The issue was food poisoning in space. If an astronaut developed food poisoning there was very little that could be done. No one was there to treat the patient, and since every member of the on board space team had a crucial part to play in the mission, a bout of food poisoning would have had a devastating effect on a mission. So NASA developed methods to minimise the risk of food poisoning in space. They (in collaboration with the Pillsbury Corporation) produced a food safety control system that they christened, Hazard Analysis Critical Control Points – this was later abbreviated to HACCP. HACCP is a system that _____ looks at a food's production process, determines the most likely places that contamination might occur, and introduces procedures to minimise contamination. This means that the food product is less likely to lead to food poisoning than if it has not gone through the HACCP process. A simple example is making a sandwich. If you make a ham sandwich for yourself you are likely to pick the ham up with your fingers and place it on to the bread, then squirt some mustard on, perhaps add a leaf or two of lettuce and a slice of tomato, before placing another slice of bread on top. Then squash the whole thing with your hand to help hold the sandwich together. There are two critical control points here where hazards (e.g. bacteria) could be introduced (Fig. 1-3).

1. Handling the ham with your fingers
2. Squashing the sandwich with your hand

To minimise the chances of food poisoning HACCP requires that you reduce, or better prevent, the possibility that the food gets contaminated at these two critical control points. To achieve this you could make sure that you washed your hands before making the sandwich, or even batter better, wear medical

Making a Sandwich

Fig. 1-3. Making a ham sandwich from a HACCP standpoint

18

gloves throughout the process. The former would be a good domestic approach, the latter would be mandatory for the commercial preparation of sandwiches.

These HACCP principles were applied in a draconian way to the Space Programme, and, as far as I am aware, were successful, no major food poisoning incidents occurred on any space flight from the USA.

In 1996 there was a serious outbreak of a new strain of bacteria (*Escherichia coli* 0157 – I'll cover this in Chapter 3) in Scotland. It resulted from contamination of a pot of gravy with uncooked meat juices. Five people died as a result of the outbreak.

Killer Bug Death

Wishaw Press, Lanarkshire, Scotland – hoarding headline in the town where the E. coli 0157 outbreak began.

A committee was set up to investigate the issue, they recommended the implementation of HACCP to prevent similar incidents occurring in the future. HACCP had been put forward for general food safety assurance. Soon legislative bodies around the world adopted HACCP and made its implementation mandatory. If you mention HACCP to many people in the food industry you will get a heart-felt groan. It can be tedious, it adds cost to food manufacturing processes, it requires a great deal of paper work, and lack of implementation can result in revocation of manufacturing licenses. But everyone knows that it is a sensible and pragmatic approach to making our food safer.

Between the 1980s and today, concern about chemicals in food has increased. As time has gone by concern has moved towards the long term effects of these chemicals. Exposure to food contaminants, or indeed natural food components, over decades at low doses might have sinister effects. Cancer-causing chemicals from many sources are a great concern to us all. Some food components and contaminants are known to cause cancer in laboratory animals. For example pesticides such as lindane have been associated by some pressure groups with particular cancers – in this case breast cancer. Others are natural, even some vitamins at high doses cause cancer (e.g. Vitamin A). On the other hand there are natural chemicals in food that protect us against cancer (e.g. Vitamin C), so it is very difficult to decide whether the carcinogens are really a problem, or whether they are balanced out by the goodies. The jury is still out on this issue as I will discuss in Chapter 7. There are other chemical contaminants that mimic female hormones (xenoestrogens). There are a myriad chemicals that fit this category, some of which might find their way into our food. Some pesticides (e.g. DDT), plasticisers (chemicals used to give plastics their characteristics), or even natural food components (e.g. genistein in Soya) are xenoestrogens. There are genuine concerns about the long-term effects of these chemicals on people, and the Newspapers have begun to report their sex changing possibilities in very graphic terms – "Gender bending chemicals in food" reported the UK Daily Mail in 2000. I will write more about these chemicals in Chapter 8.

In 1986, the face of food safety, and the public hysteria surrounding food safety issues, was changed with the first case of Mad Cow Disease (Bovine Spongiform Encephalopathy – BSE) in England. Food regulators' worst fears were realised when science proved that this unprecedented cattle disease could cause new variant Creutzfeldt Jacob Disease (nvCJD) in people who ate infected beef. Now a single case of BSE in a country leads to other nations banning beef imports from that country (e.g. Canada 2003) despite the infinitesimally tiny risk of the imported beef leading to disease in the importing country. This is sheer stupidity driven by public perception fuelled by the press... but more of this in Chapter 5.

Why we should all give up beef?
"How can the public make sense of the row about BSE when politicians and experts cannot agree?"

From Professor I C Shaw
Sir: A thought for the Prime Minister, who said today that there is no scientific evidence that BSE and CJD are linked. Before gravity was demonstrated, there was no scienctificscientific evidence for its existence.
Yours faithfully,
I.C. SHAW
Hambleton, Lancashire
7 December

Headline and a letter from the UK's Independent newspaper, 7 December 1995

As a result of the furore that surrounded the BSE saga, beef markets collapsed, people changed their diets, governments legislated, and many people were suspicious about the safety of their food. So is food really that risky? Read on ...

We won't swallow and more lies about food

Headline from the UK's Independent newspaper, 31 January 1997)

2
Food is Just One of Life's Risks

2   Food is Just One of Life's Risks

What is Risk?

Everything that we do is risky, even sitting in your chair at home reading this book has risk associated with it. The roof of your house might fall in and kill you, you might have an earthquake, or a meteor might hit your house. These risks are infinitesimally small, but they are nevertheless real risks. Despite this most of us would pay little attention to such minute risks. It is important that we keep risks in perspective so that we can decide what we need to worry about. As I discussed in Chapter 1, eating is a risky business – but how big is that risk? And how much should we worry about it?

Before we can decide whether we should be overly concerned about the risks associated with eating we need to understand what risk is. The Oxford English Dictionary includes in its definition of risk:

"exposure to mischance ... exposed to danger ... expose to chance of injury or loss."

It also uses the word *hazard* to describe risk. In scientific parlance *risk* and *hazard* are distinct and MUST not be confused. Hazard relates to the intrinsic deleterious properties of a chemical, living organism, or physical effect (e.g. radioactivity). So, for example, potassium cyanide has a very high hazard. Hazard can be measured in animal experiments by determining how much of a chemical is needed to kill a test species (usually expressed as the dose necessary to kill 50% of a test population of animals – LD50), or determining how much of the chemical is needed to cause a measurable effect (e.g. the change in the level of a hormone in an animal given a dose of the test chemical) – this is termed No Observable Effect Level (NOEL) and has now replaced LD50 tests because it is a far more humane way of checking toxicity. The LD50 for potassium cyanide by oral administration to rats is 10 mg/kg body weight (i.e. it takes a dose of 10 mg of potassium cyanide per Kg body weight of the rat – an average adult rat weighs about 500 g, so the lethal dose would be about 5 mg – this is the weight of a few grains of salt). Clearly potassium cyanide is extremely toxic. If we extrapolate the lethal dose in the rat to humans (average weight of a human is 70 kg), it would take about 70 mg to kill one of us.

Risk is related to exposure to the hazard and is a far more important way of expressing danger than talking about hazard. Unfortunately some politicians and action groups have not yet grasped this concept! If you have a bottle of potassium cyanide on the table in front of you it has an incredibly high hazard, but the risk to you is tiny because you are unlikely to eat it. If you don't open the bottle your exposure is zero and therefore the risk is zero.

Risk = Hazarad × Exposure

This simple equation is the basis of the science of risk.

In years gone by – before the authorities got worried about even the smallest risks – I used to demonstrate hazard and risk to my students by taking a bottle of potassium cyanide, weighing out 1 g and dissolving this into a litre of water. The concentration of that solution was 1000 mg/l. If I had drunk 70 ml I probably would have died. That was far too great a risk to take. So I took 1 ml of the solution and made it up to 1 l with water. This gave a solution with a potassium cyanide concentration of 1 mg/l. I happily drank a small glass full of the solution in the knowledge that I would have to drink 70 litres to kill me. This is an excellent illustration of hazard and risk. In the final example the cyanide is incredibly high hazard, but very low risk. And what's more I'm here to tell you the story!

An even simpler illustration of risk and hazard involves three vicious hungry animals in three cages, all have not been fed for a few days and so are very hungry (Fig. 2-1).

The animals represent equal hazards – they are all hungry and would eat you if given the chance. The risk from the animal in cage 1 is near to zero because barring some weird quirk of nature that resulted in the cage door opening, it would not be possible for the animal to get out to eat you, i.e. your exposure to the animal is zero. The chance of exposure to the animal in cage 2 is higher because its cage door is unlocked, therefore the

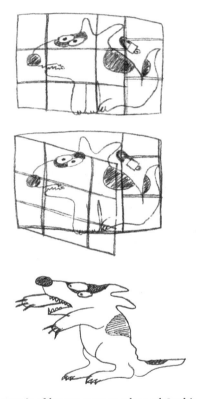

Fig. 2-1. Risk is determined by exposure to a hazard. In this example the hazards are identical – 3 vicious animals, but the risks are different. The first animal is locked in his cage therefore exposure is not possible – risk = 0; the second animal's cage door is unlocked therefore exposure is possible – risk = moderate; the third animal is not caged therefore exposure is inevitable – risk = high

risk is greater. The risk from the animal in cage 3 is enormous. His cage is open and he is free to leave and eat you.

Risk is a measure of the magnitude of the effect of our exposure to hazards. Crossing the road is a risk. We look both ways before crossing in order to minimise the risk, but still people get injured or killed crossing the road. It's the same with other risks, including the risks associated with eating. We attempt to minimise our exposure to hazards and so minimise the risk. We maintain good hygiene in our kitchens to minimise our exposure to disease causing bugs on our food. In industry they apply HACCP to minimise food risks. Our survival instinct makes us avoid risk, but it also often makes us exaggerate risk in our own minds – if it appears worse than it is you are more likely to avoid it! There is a difference between "real" risk (i.e. assessed using the risk equation) and assumed or perceived risk. Rare risks are usually perceived as worse than every day risks. So to most people crossing the road is safe – we think nothing about it, but flying in an aeroplane is more worrying. In reality you are much more likely to be killed crossing the road than you are in an air accident. The press helps us to perceive risks. They rarely publish articles about people being killed crossing the road, or in road traffic accidents because this is far too common to be interesting. On the other hand a hundred people killed in an air crash is front page news. So our perception of the risks is fuelled by the media.

BSE Risk

The BSE saga in the UK was a disaster for the beef industry, farmers were put out of work, and some even committed suicide. These were terrible times. The press reported every sordid event. They filled columns, and emblazoned headlines on numerous front pages of quality newspapers. You can understand why the British, and later the rest of the world, were terrified of BSE. They thought that a single mouthful of beef would give them nvCJD and that they would die a horrible death. The newspapers were reluctant to publish the real risk statistics. The risk of getting nvCJD was, and is, exceptionally small. Many millions of times less likely than dying in a car accident. But then this would not have sold newspapers. This might be a rather cynical view, but I suspect it is at least in part true. In the broad array of life's risks it seemed strange to me, as a risk scientist, that people were unhappy to eat beef, but did not give driving their car a second thought.

For us to respond appropriately to risks we must rank them. This helps to decide which ones to act upon; which ones are most important. Clearly BSE was important and had to be acted upon. This is unquestionable. It was an added life risk that we could minimise by legislation to prevent it re-occurring (I'll discuss this fully in Chapter 5).

Is Smoking an Acceptable Risk?

There are many other risks that people seem to accept, but why? Smoking is a good example. This is an addiction to a drug (nicotine) that is delivered in a rather bizarre way (i.e. in smoke) that simultaneously delivers highly hazardous chemicals (cancer-causing tars). The health risks associated with smoking are bronchitis, heart disease, asthma, and lung cancer. It seems strange that anyone would take this risk. In fact it seems utterly stupid! However, there is another side to the risk equation – benefit. Some people enjoy smoking. The benefit of smoking is enjoyment, pleasure. It is not possible to understand risk until we bring into the equation benefit. If the risk outweighs the ben-

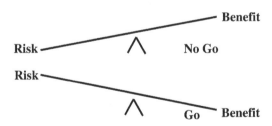

Fig. 2-2. Whether risk is acceptable depends upon benefit. If benefit outweighs risk the situation is acceptable. If risk outweighs benefit it is unacceptable

efit, the risk is unacceptable. If the benefit outweighs the risk the risk is acceptable. To smokers the benefit of a cigarette far outweighs its risk ... and so they continue to smoke despite the potential effects upon their health (Fig. 2-2).

Food Risks

Food risks can be assessed in just the same way. To many Japanese people Fugu Fish is a great delicacy despite its potential toxicity. Fugu Fish contains tetrodotoxin (LD50 = 10 µg/Kg) which is intensely toxic – just 0.07 mg is likely to kill a human, this is equivalent to a small fragment of a grain of salt. The tetrodotoxin is present in the bile of the fish and when the chef prepares the fish meat for consumption he allows a tiny drop of the bile to contaminate the flesh. He does this because the toxin is a nerve poison and causes the consumer of the fish to get a tingly sensation on their lips. This is the benefit that the consumer gets to set against the risk of being killed by the toxin. The diners in Fugu Fish restaurants must really trust the chef. One mistake and they're dead. Clearly the Japanese government does not fully trust the chefs because they have recently introduced legislation to control the preparation of Fugu Fish so minimising the risk of harming its consumers – chefs now have to be trained

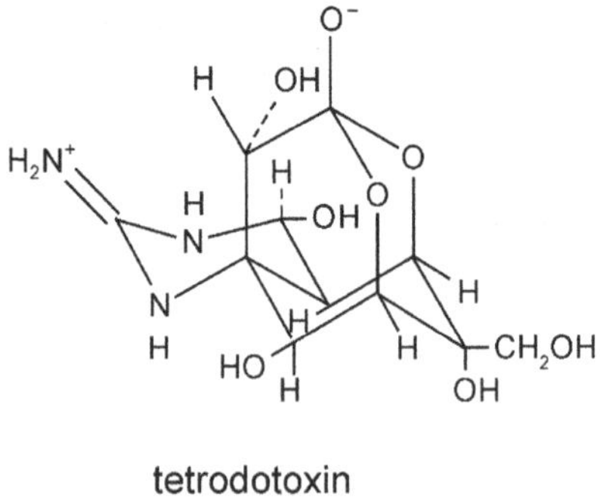

tetrodotoxin

Fig. 2-3. The Complex Molecular Structure of the Puffer Fish's Deadly Toxin, Tetrodotoxin

and pass an examination before they can prepare Fugu sashimi. They did this because every year about 8 people die of tetrodotoxin poisoning in Japan and this risk was not considered acceptable by the Japanese Ministry of Health & Welfare – quite right too! (Fig. 2-3).

Food associated risks are low compared to many of life's risks. Indeed the greatest risk associated with food is going to the shop to buy it. You are much more likely to get killed or injured on the way to the shop than you are to be harmed by the food that you buy.

HIGH RISK
Getting run over on the way to the shop
Choking on a Brussels sprout
Natural toxins
Pesticide residues
LOW RISK

Taken from "The risks of eating" Shaw IC (1999) Pesticides in Food. In: Brooks GT, Roberts TR (eds) Pesticide Chemistry and Bioscience. RSC, London

Continuing this comparison of food-associated risk with other life risks, it is possible to give risks numerical values by using the *risk = hazard × exposure* equation. These can be plotted on a graph to allow a comparison of risks to be made (Fig. 2-4).

The food-related illnesses are right at the bottom left hand corner of the graph, i.e. they are the lowest risks. So why are we worried about them?

I suppose the best answer to this question is that we worry because we feel that food should be safe. But this is relative, and in the context of life's daily risks that we accept without question, food is indeed very safe. Despite this, I'm pleased that we worry about food risks because it is possible to minimise them and so reduce life's risk burden upon us. For example, there is a world-wide problem with a relatively common food contaminant, the bacterium *Campylobacter jejuni*; it probably kills thousands of people around the world each year. Many of these will be in third world countries where food regulations are

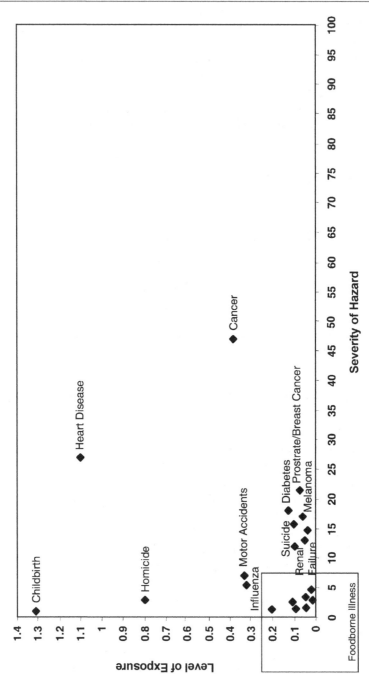

Fig. 2-4. The risks of being a New Zealander, showing that food poses one of the lowest risks (prepared in 2002 by Eva Harris, one of my summer vacation students)

scant or non-existent. However, in developed countries such as New Zealand, campylobacteriosis is still a significant food issue. Each year 1 or 2 people die in New Zealand of campylobacteriosis – in a population of only 4 million this is a significant number – and this is just the tip of an enormous iceberg, because for every death there are hundreds of cases of illness (I will deal with this more fully in Chapter 3). The question, of course, is "can we reduce the number of Campy infections and so reduce this risk?" The answer is almost certainly yes. Chicken has been suggested as the major source of Campy (the poultry industry disagree with this, and more work needs to be done to prove it one way or the other), so risk from Campy could be reduced by treating chickens to kill any Campy that they might be harbouring. In Iceland all chickens are frozen (this seems rather appropriate!) before selling for human consumption; freezing kills Campy – what a good idea. This (if it works, and we have every reason to believe that it will) is a very definitive intervention. However despite its scientific foundation, we might not like only being offered frozen chickens. Fresh chicken tastes better. Some of us might be prepared to accept the Campy risk for the better tasting fresh chicken – I certainly would. Choice is important, and for this reason most governments will not intervene in a draconian way unless the risk that they are reducing is enormous. Campy risk is not enormous. A better way of dealing with it is to educate the public about Campy. How can we reduce our chances of catching it? Campy is destroyed by heat, so proper cooking removes the problem. This is a far better solution than reducing the consumer's choice. Public education is the way forward. Tell people about hazards, and tell them how to reduce risks. They can then make up their own minds. That is exactly what this book is trying to do (Fig. 2-5).

If we identify hazards, assess risks based on exposure to the hazards, and then communicate exposure routes and means of avoiding them to the consumer, changed behaviour might result in reduced exposure and reduced risk. The hazard remains in place but its effect on the consumer is minimised. This is a cheaper option for regulators than trying to eliminate the hazard. This process is termed risk management. It relies on good consumer communication strategies.

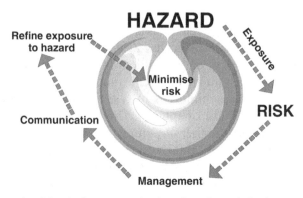

Fig. 2-5. The risk spiral – communication about hazards leads to control of exposure and minimises risk (from Shaw, IC (2002) Making Food Safe to Eat, Food Technology New Zealand, November)

To illustrate this, I will return to the Campy example. Hazard elimination could involve selling only frozen chicken. On the other hand, risk management might involve alerting the consumer to the need to cook chicken well to kill Campy, or to the problems of cross contamination when cooking utensils are used to handle raw chicken and then are used to serve up the cooked meat. Both of these pieces of information will help the consumer to modify their behaviour so reducing exposure to this potentially lethal bacterium. It is simple, don't use the same tongs to handle raw and cooked chicken and so reduce the risk of getting campylobacteriosis. It might seem simple, but it can't be effective unless we communicate the hazards and how to minimise exposure to them to the consumer.

Risk Perception

The discussion above relates to the "real" risks associated with living. Crossing the road, meeting hungry lions, eating barbecued chicken, etc. However there is another very significant force at play when we consider risk – PERCEPTION. I can calculate risk and express it quantitatively, for example the risk of suffering from cancer in the UK is 1 in 220/year – if you live in the UK you have a 1 in 220 chance of getting cancer (i.e. there are 267,000 newly diagnosed cases of cancer in the UK's population of 58,789,194 each year). As you get older the risk goes up, because cancer is a disease of advancing years. Did you expect the risk to be this low? What you thought the risk was is the perceived risk. Usually we perceive common risks as being low and rare risks as being high. So, it is likely that you thought the risk of getting cancer was higher than it is.

An everyday example illustrates this well. Try putting the following activities in risk order:

- Flying in an aeroplane
- Travelling in a car
- Dying from nvCJD from beef in the UK

There have been many studies on the perception of risk that show that many people would put flying in an aeroplane at the top of travel risk. So, let's look at the statistics. I'll use UK statistics to illustrate my point, but stats from all developed countries would show the same trend, Table 1.

The risk of death from car accidents is very much higher than the risk of death in an air accident. This only relates to one year in the UK, but shows clearly that no air-related deaths resulted from 73×10^{11} km travelled, but 1,687 people died on the roads over a total of 6.5×10^{11} km travelled (i.e. 91% less than air travel distance).

UK domestic travel/death statistics (1999). Data from National Statistics Online: http://www.statistics.gov.uk		
Transport	Total distance travelled, km/year	Deaths
Car	6.5×10^{11}*	1,687
Air	73×10^{11}	0

* An average person in the UK travels 10,904 km/year, therefore assuming that everyone (population = 60 million) in the UK travels by car, the total distance travelled by car is $10,904 \times 60 \times 10^6 = 6.5 \times 10^{11}$ km.

Activity	Risk of death in the UK in 1999
Travelling in a car	1 in 35,714
Dying from nvCJD from Beef in the UK	1 in 4,000,000
Flying in an aeroplane	0

Assuming that everyone in the UK travels about the same distance each year by car, this gives a risk of dying in a car accident of 1 in 35,714/year. How does this compare with the risk of contracting nvCJD from beef in the UK? A lot of assumptions need to be made to do this calculation (e.g. that everyone eats beef – of course they don't), but the risk comes out to about 1 in 4,000,000 in 1999 (i.e. there were 15 newly diagnosed cases in 1999 out of a population of 60 million). This is not a good risk calculation because the "incubation" period for nvCJD is long – up to 12 years – and so occurrence of the disease in 1999 relates to exposure years ago. Nevertheless it gives us an idea of the magnitude of the risk, and it is very low; much lower than being killed in a car accident.

Using these risk values we can put the three activities that were discussed above in risk order (i.e. rank them), Fig. 2-2.

Most people wouldn't think twice about driving their car, but are likely to worry a little about air travel and eating beef. But they are wrong; they should worry very much more about driving their car than either of the other activities. Their per-

Perceived Risk

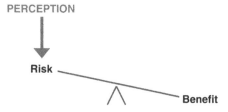

Fig. 2-6. The risk/benefit balance showing how perception adds weight to risk

ception of the BSE risk is far greater than reality – probably fuelled by an over zealous news media.

Perception often adds weight to the risk side of the risk/benefit balance, and therefore means that the risk seems to be greater and so is less likely to be outweighed by the benefit. For example, the threat of contracting nvCJD from beef in the UK meant that many people stopped eating beef even though they enjoyed it. The risk is nvCJD, the benefit is enjoyment, but their perception of the risk was greater than reality and therefore it outweighed the benefit in their minds (Fig. 2-6).

Paracelsus – the Grandfather of Risk

Risk science is not new, in fact a German scientist Phillipus Aureolus Theophrastus Bombastus von Hohenheim Paracelsus was its first, as far as we know, protagonist. Paracelsus was born in Switzerland in 1493, and died in 1541. He is famous for his risk philosophy discussed in his Four Treatises (Fig. 2-7):

Alle Dinge sind Gift   All things are poisons
Und nichts ohne Gift   There is nothing which is not a poison
Allein die Dosis macht,   It is the dose
Dass ein Ding   Which makes a thing safe
Kein Gift ist

Roughly translated from Paracelsian philosophy this means, a little bit will do you no harm. Remember our discussion on cyanide where a tiny dose, even though its hazard is very high, will cause no ill effects – cyanide is poisonous, but a small dose won't hurt you.

Is There a Price on Risk?

The benefit side of the risk/benefit balance can have more than one component. For example it could include pleasure and price. If one of these changes it can tip the balance in favour of benefit.

Fig. 2-7. Theophrastus von Hohenheim Paracelsus (1493–1541) (reproduced from http://www.mhiz.unizh.ch/Paracelsus.html by kind permission of Dr. Urs Leo Gantenbein)

During the BSE epidemic in the UK there were significant fluctuations in the sale of beef. When it was suspected that BSE might cause disease in humans who consumed contaminated beef, sales of beef dropped off sharply. When it was confirmed that nvCJD resulted from consuming BSE beef, sales fell drastically and the UK beef industry collapsed. However the price drop that resulted from the decreased demand for beef meant that some people thought that beef was such good value that they were prepared to risk contracting nvCJD and eat beef anyway. Therefore sales of beef rose shortly after the price hit rock bottom. The power of price and enjoyment of eating beef outweighed the perceived risk of nvCJD. In this case the risk was acceptable at £1.95/lb! (Fig 2-8).

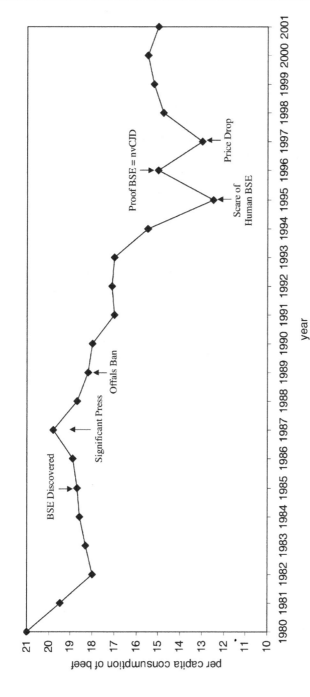

Fig. 2-8. Is there a price on risk? The changing price of UK beef driven by the Mad Cow saga and its effect on consumption. In 1997 the price got so low that people "forgot" their concerns about eating Mad Cow beef and stocked their freezers this increased demand drove the price up again (data from UK Agriculture Committee (2002))

Is Risk Increasing?

On the face of it, it looks as though life is getting more risky as time passes. More people die of cancer now than 100 years ago. This is partly because cancer diagnosis was rather hit and miss a century ago, and partly because cancer is a disease of old age and people are living longer now and therefore the susceptible population is greater. The fact that people are living longer means that risk must be decreasing. This is largely because medicine is better now than in years gone by. The risks associated with many diseases have reduced drastically. Tuberculosis (TB) is an excellent example. Just 75 years ago, before the introduction of sulphonamides, many people died of TB, now it is relatively easy to treat and therefore the death rate in the developed world from TB is much lower than pre-sulphonamides (Fig. 2-9).

Therefore, in general, life's risks are decreasing because of better medical care and medicines. This is why Queen Elizabeth II sends more congratulatory telegrams to her centenarian subjects than Queen Elizabeth I might have. Associated with this decreased risk of living there is a concomitant increase in some risks. The decreased risk is associated with intellectual and technological advances, e.g. medicine. With these come extra risks. The invention of the motor car is an obvious example. Despite all of this, we live longer now than we did 100 years ago and therefore the risks are being managed effectively to promote a longer life.

Amidst life's risks are food related risks. We have discussed some of these above – they are incredibly low in the context of many of our other daily risks. But the question is, are they increasing? A quick look back at Chapter 1 would persuade anyone that food related risks have diminished over the centuries. However we need to look more closely at the risks over the past few decades. We hear so much more about food related illness now than in decades gone by. We go to exotic holiday destinations and return with gastric upsets. Travel agents even warn us not to eat certain foods in some of these tropical havens. A few days in India or Indonesia without a very sensible approach to eating will prove my point. I speak here from bitter experience! There is absolutely no doubt that eating in developing countries

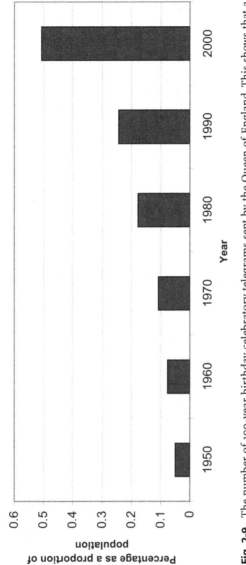

Fig. 2-9. The number of 100-year birthday celebratory telegrams sent by the Queen of England. This shows that a greater proportion of the population are living beyond 100 years so life must be getting less risky

is far more risky that eating in the developed world. Most developing countries do not have good food legislation. This is way down their list of priorities – feeding their people and combating terrible diseases is rightly above food safety. So if we travel to these countries we apply our own risk management procedures – don't eat uncooked, unpeeled vegetables, watch food that might be reheated, don't drink water unless it is bottled, don't have ice in drinks, etc. This minimises our exposure to microbiological hazards associated with food and water.

But what about the so-called developed world? How have food-related illnesses changed in the past few decades? The USA has an excellent organisation, the Centers for Disease Control and Prevention (CDC) that collects and collates (amongst a myriad other things) data on food poisoning. Since the USA is a good example of a developed country from the point of view of food safety issues, I will illustrate the changes in food-born illness over the past decades with information from CDC.

In the USA, about 76,000,000 food-related illnesses occur each year (this includes multiple incidents in individuals), of these 325,000 (0.43% of cases) result in hospitalisation, and there are 5,000 deaths (0.007% of cases). *Salmonella, Listeria* and *Toxoplasma* (a parasite, found in food, that causes neurological disease) account for 1,500 deaths per year. Looking at just one of these food-related infections, listeriosis, in the USA over a 7-year period shows a clear downward trend (Fig. 2-10).

Other diseases (campylobacteriosis and yersiniosis) show the same trend. This suggests that food-related illness risk is going down. But if we look further, other bacterial diseases associated with food are showing a gentle upward trend (Fig. 2-11).

This is a "swings and roundabouts" situation. Some food-related diseases are increasing, while others are on the way down.

Chemical Risks

All of our discussion so far has been about microbiological food risk. This is because far more is known about the diseases that

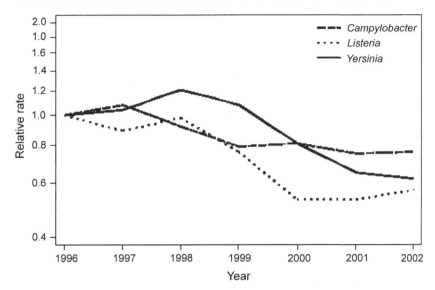

Fig. 2-10. Declining numbers of listeriosis, campylobacteriosis and yersiniosis cases in the USA (data from Mead et al. (2003), Food-related illness and death in the United States http:www.cdc.gov/ncidod/eid /vol5no5 /mead/htm)

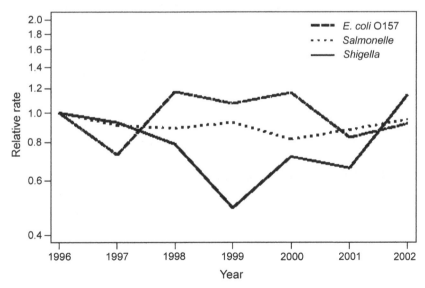

Fig. 2-11. Declining numbers of *E. coli*, *Salmonella* and *Shigella* cases in the USA (data from Mead et al. (2003), Food-related illness and death in the United States http:www. cdc.gov/ncidod/eid/vol5no5/mead/htm)

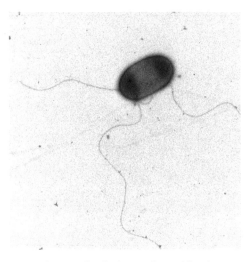

Fig. 2-12. Electron micrograph of *Listeria* [magnification approx 35,000X] (kindly provided by Phillipa Rhodes)

microbes cause. The symptoms of *Listeria* infection (i.e. listeriosis) are very well known; any doctor would recognise them and could prove their cause by taking a faeces sample and getting the lab to culture it to show the presence of *Listeria* (Fig. 2-12).

Chemical contaminants are very different. Most food-related bacterial and viral diseases are acute – i.e. the disease associated with the virus or bacterium occurs a day or two – or at most a few weeks after infection. It is not so simple for chemical contaminants. Indeed, even if a particular chemical contaminant could cause an acute effect, it is extremely unlikely that the levels in food would be sufficiently high to result in acute toxicity (i.e. exposure to the acute hazard is too low to result in an acute risk). For example, organophosphorus pesticide (OP) residues – from the use of OPs in crop protection or as vet medicines – in food would never result in the acute neurological effects (shaking, salivation, death) associated with very high doses of OPs. This is because residues in food are at exceptionally low levels (i.e. of the order of parts per million [ppm], or mg/kg) compared with the doses that would cause acute toxic effects.

Pesticides

Data from the UK's pesticide monitoring schemes show that some vegetables have very low residues of an OP called pirimiphos-methyl – used as an insecticide in crop production and grain storage. For example, a single pirimiphos-methyl level found in 1999 in bread was 0.1 mg/kg. If you assume that someone eats 250 g of bread in a day – this is a worst case scenario because it wrongly assumes that all bread contains pyrimiphos-methyl residues at 0.1 mg/kg (most would contain no residues at all), the daily dose of pyrimiphos-methyl would be 0.025 mg. Assuming that the person weighs 60 kg, their daily intake would be 0.0004 mg/kg body weight. The maximum dose that would not result in toxicity to a person if they ate it every day for their entire life (this is termed the Acceptable Daily Intake – ADI) is 0.01 mg/kg, and this includes a safety factor of 1,000. The bread that the UK found to contain residues of pyrimiphos-methyl would lead to a dose that is only 4% of the ADI and therefore would not result in toxicity; the risk is therefore very low indeed.

The same sort of calculation can be made for most pesticide residues in food. Indeed this argument has often been used to show that residues are of no health concern to the consumer. But what effect might the complex cocktail of pesticides that we eat in a lifetime have on our health? We don't know, and it is extremely difficult to predict or determine such effects because the cocktail changes with time as new pesticides are introduced and old ones phased out. So we certainly cannot say that pesticide residues will have no effect, but we can say that they are unlikely to have an effect based on the toxicity of individual pesticides in our food.

A simple approach to determining the possible additive effects, as a means of assessing risk, of pesticide residues is to look at a class of pesticides whose members all exert their effects by the same biological process (mechanism). If we determine risk based on the individual pesticides this gives us a low risk assessment, but since they are all acting by the same mechanism we should add up their effects to give an overall effect of exposure to the class of pesticides.

The OPs are an excellent example. They all kill insects by preventing nerve impulses being generated by inhibiting a specific enzyme (acetylcholinesterase – AChE) in the nervous system (see Chapter 7). Similarly their toxic effects on people are by the same mechanism. So if you get small doses of Propetamphos, pyrimiphos-methyl, Diazinon and a large number of other OPs, determining risk by looking at the individual OP intakes will not give a realistic assessment of the risk. The best way would be to add the concentrations together (or better, add together a measure of their relative biological activity) and determine an additive effect. It might be that each of the pesticide intakes is just below the ADI – this is unlikely, but will be used here as an example. Adding them together would exceed the ADI, and might result in harm to the consumer. I will discuss this again in Chapter 7, but for now it is important that we remember that chemical risks might appear lower than they really are because it is very difficult to assess them properly.

Newspapers Often Exaggerate Risk

Newspapers often exaggerate the risks associated with chemical residues in food and therefore increase the consumer's perception of risk. For this reason many people rate the risks of pesticides in their food higher than microbiological risk. This is simply wrong.

Your typical lettuce.... after 11 doses of pesticide

The front page newspaper headline from the UK's Guardian newspaper on 16 September 1999 reporting the "leaked" annual report of the Working Party on Pesticide Residues – beneath the headline was a picture of an apparently normal lettuce. The article was accompanied by a very funny cartoon showing a woman off to do her shopping wearing protective clothing! The article quoted statistics on pesticide residues frequency, but did not highlight their levels; it focused on hazard rather than risk – this is misleading.

A great deal of this over perception of the risk of pesticides is due to adverse press coverage. It is interesting that in a survey of experts (in this case toxicologists) and lay people (i.e. non-scientists). The experts rated pesticides as less toxic than did the lay group:

Statement: Residents of a small community observed that several malformed children had been born there during each of the past few years. The town is in a region where agricultural pesticides have been used during the past decade. It is very likely that these pesticides were the cause of the malformations.

Responses (%):	Strongly disagree	Disagree	Agree	Strongly agree	Don't know
Toxicologists	22.2	59.3	4.3	1.2	13.0
Lay people	3.9	23.4	39.5	9.0	24.2

Taken form Paul Slovic's The Perception of Risk, Earthscan Publications Ltd., London, 2000.

It is interesting that the experts do not regard pesticides as being the cause of the town's malformed children, whereas the lay people blame the pesticides. This is an important illustration that perception depends on knowledge. This is why we teach our children that cars are dangerous, in the hope that they will err on the side of safety when crossing the road. "Err on the side of safety" means over assess the risk. This is exactly what the lay group did in the above example.

In conclusion, risks are not always what you think they are. If the risk that you are contemplating is common place (e.g. smoking) you are likely to under estimate it. Whereas if the risk is rare (e.g. death from vaccination) you are more likely to over rate it. When students were asked to assess the risk of smoking-related deaths per year in the USA, they came up with 2,400 – the real value is 150,000. On the other hand they were asked to assess deaths from skiing, they thought 72 – the real figure is 18. I bet more of them smoke than ski!

47

Paracelsus should have the last word. *All things are poisons, there is nothing that is not a poison, it is the dose that makes a thing safe,* or in other words, everything is safe unless you take too much of it!

3
Bacteria in Food – Good or Bad?

3   Bacteria in Food – Good or Bad?

Bacteria are everywhere. They coat our bodies, they live in our gut, up our noses, in our throats. They are in soil, the water we drink, river and pond water, the sea…everywhere. Most of them are harmless, we don't give a second thought to licking our finger, kissing our loved one, or eating a juicy apple straight from the tree. With all of these activities we consume bacteria. We don't think twice about doing these things because we don't expect to be harmed by the experience. If we assessed the risks, they would be low. But every now and again we might take in a bacterium that will lead to a disease – a pathogen. If you kiss someone with a throat infection, you will probably get a sore throat too. The throat infection might be caused by pathogenic bacteria, probably *Streptococcus* species – hence 'strep throat'.

Before we move into the world of food bacteria, we'll have a closer look at bacteria in general.

What Are Bacteria?

Bacteria are microscopic plants. They are single celled (although some might grow as long filaments, but even then each cell functions separately), have no nucleus (termed prokaryotes – meaning before nuclei in Greek), and have a rigid cell wall. They were amongst the first cells to evolve in the Primordial soup of 4,000,000,000 years ago. They have changed little over this long time period.

Bacteria can live in the most amazing diversity of environments. They are found living around volcanic spouts in the deepest oceans. In this environment they are subjected to enormous water pressures, and temperatures in excess of 100°C. They live at the Arctic and Antarctic polar regions at sub-zero temperatures, they live at extremes of pH, and can tolerate concentrations of toxic chemicals that no other species would survive in. Some even use cyanide as their basic metabolic energy source, others use sulphur instead of carbon – these are the bacteria that produce hydrogen sulphide (bad egg smell) in the mud at the bottoms of ponds, some don't even require oxygen to survive (anaerobic bacteria).

Some bacteria live a static life attached to their substrate, others move with the flow of the water that they live in, some have hundreds of minute beating hairs (cilia) to help them to move, and others have long whip-like structures (flagella) that beat and propel the cell along (Fig. 3-1).

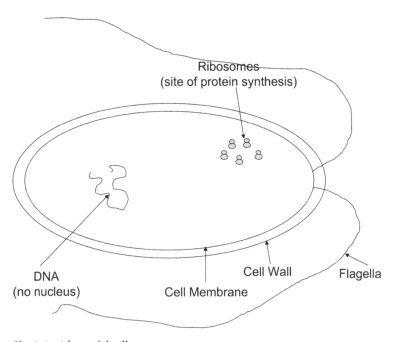

Ribosomes
(site of protein synthesis)

DNA
(no nucleus)

Cell Membrane

Cell Wall

Flagella

Fig. 3-1. A bacterial cell

As mentioned above, bacteria are microscopic. The Dutch scientist Antony van Leeuwenhoek (1632–1723) was the first to see a bacterium through his newly invented microscope. They are only about 5 μm (5 millionths of a metre) long. To see them they must be magnified at least 100 times when they appear as barely visible dots or rods depending on the type of bacterium. Today we have incredibly sophisticated microscopes that use electrons instead of light (i.e. the electron microscope) and are able to magnify up to 100,000 times, thus allowing us to see the intricate structures of bacteria easily.

Most bacteria are good. They are an integral part of our environment, for example they are important in breaking down waste. Without these bacteria we would disappear under a great heap of our own excrement. The compost heap in the garden turns from a pile of lawn mowings, kitchen waste and garden prunings, into a dark, loamy nutrient-rich compost aided by bacteria. The landfill sites in which we bury our domestic waste can eventually be turned back into farm land, or built on because bacteria breakdown the enormous amount of apparently indestructible stuff that we throw out.

Harmful Bacteria – Pathogens

Unfortunately a relatively small number of bacterial species are able to cause disease in people, other animals and plants. These are the bacterial pathogens, and they are a significant problem to farmers, vets and doctors alike.

Louis Pasteur and Pasteurisation

Louis Pasteur, the French microbiologist (born 1822, died 1895), demonstrated that bacteria can cause disease – the Germ Theory of Disease, and that they can be killed by heat (hence pasteurisation). This was a great medical discovery – before then there was little prospect of treating many diseases because doctors did not know what caused them.

Pasteur's name lives on in pasteurisation, the process of killing bacteria by heating liquids or solids (e.g. milk) to a sufficiently high temperature to kill the pathogens. The Pateurisation of milk involves heating to 63°C for a 30 minutes, or to 72°C for 15 seconds. The importance of pasteurisation of food is that it can kill bacteria without significantly changing the taste of the food. Alternatively sterilisation can be used, this involves boiling liquids, but it usually changes the flavour of the food very significantly. When I was a boy the milkman in my home city of Birmingham in the UK Midlands would deliver both sterilised and pasteurised milk. Even as a young lad I hated the taste of sterilised milk – it tasted very different to pasteurised milk. Also the pasteurisation process is gentle on the nutritional components of the food. For example, sterilisation of milk denatures the protein components (e.g. casein, milk's major protein), so changing their absorption and metabolism by the body and perhaps their nutritional value (Fig. 3-2).

Fig. 3-2. A carton of pasteurised milk – a common sight in a modern refrigerator

Ultra High Temperature (UHT) Pasteurisation

There are several other higher temperature pasteurisation processes used nowadays. If you look at long-life dairy products you will see that these are the processes used to kill even the most tenacious bacteria and so give the product a very long shelf-life. Ultra high temperature (UHT) pasteurisation involves heating the product (e.g. milk) to between 138 and 150°C for a few seconds. Providing the milk is packed in sterile, sealed containers it will keep for moths without refrigeration.

These are good methods to kill both bacteria that cause food to go off (i.e. spoilage bacteria), and bacterial pathogens. This means that pasteurised products last longer and are safer than their "raw" counterparts.

I mentioned flavour above. Any heat treatment, even the gentle heat of pasteurisation or the short blast of heat used in UHT pasteurisation, changes the flavour of the product. For this reason there dissention in gourmet circles about the process. Some prefer to take the risk of being infected to taste the food as it should taste. Milk is a good example. Many people say that raw milk tastes better than pasteurised milk and will go to great lengths to obtain the unadulterated product – Queen Elizabeth II is said to enjoy un-pasteurised milk and might even have influenced the UK legislators when they tried to ban un-pasteurised milk in the UK.

Cheese

Un-pasteurised cheeses are an even greater issue. No self-respecting Frenchman would eat a bland pasteurised cheese! There is absolutely no doubt that the flavour of raw cheeses are more complex (and in my opinion better) than their sanitised counterparts. The reason for this difference in taste is that if raw milk is used in the cheese making process, and the cheese is not pasteurised, not only the bacteria added to commence the fermentation process are present, but a myriad other bacteria, yeast, and fungi take up residence in, and on, the cheese. These creatures all excrete flavouring chemicals into the cheese, giving it its unique complex taste – my mouth is watering just thinking about it! This is why cheeses from different manufacturers that use very similar processes can taste very different – their factories infect the cheese with different microorganisms and together they impart their own unique flavour.

So, what is the risk of eating un-pasteurised dairy products? pasteurisation was originally applied to milk to kill *Mycobacterium tuberculosis* (causes tuberculosis), and is still important today for the same reason. However the infection rate in cattle is now much lower, and therefore the chances of catching tuberculosis from even un-pasteurised milk is very low indeed. However, in my opinion, the difference in taste between pasteurised and un-pasteurised milk is so small, that the benefit is not worth the risk – so, I would always drink pasteurised milk. However, to my mind, cheeses are a very different matter. I can think of nothing better than a gooey, blue veined, creamy, rich, smelly, completely yummy St Augur cheese from one of those beautiful French cheese shops piled floor to ceiling with maturing cheeses, and smelling like a pair of socks badly in need of a trip to the laundrette. The benefit of the enormous difference in taste makes pasteurisation of cheese a crime in my opinion. Many people, especially the Legislators, would disagree with me though… perhaps they should remember the risk of crossing the road!

Good Bugs

As outlined above there are many more good than bad bacteria. Good bacteria colonise food and prevent bad (pathogenic) bacteria growing – they simply outgrow them, or some produce toxic chemicals to prevent competing micro-organisms growing. This is an important aspect of food ecology that our desire to live in a sterile world has overlooked. Indeed, sterile food has the potential to be more harmful that food supporting a good, natural microbial ecology because if a "sterile" food is inoculated with a pathogen it will grow unchecked.

Some bacteria produce chemicals that help to preserve food. For example *Lactobacilli* produce lactic acid. Lactic acid

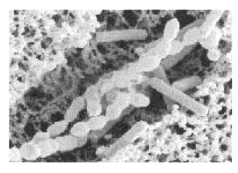

Fig. 3-3. Yogurt under the microscope showing the bacteria (*Streptococci and Lactobacilli*) responsible for its thick texture and flavour (from http://www.customprobiotics.com/yogurt_culture.htm by permission of Custom Probiotics, California)

makes milk curdle and is the first step in cheese making, but it also has good preserving properties. Salami is lactic fermented meat, the high concentrations of lactic acid preserve the meat and add a piquancy to its flavour. Lactic acid is now manufactured 'artificially' and added to some foods as a preservative – it is E270 in the European Union E-number labelling system.

Yogurt is milk fermented by *Streptococcus thermophilus*, *Lactobacillus bulgaricus* or sometimes *L. acidophilus*. The *Lactobacilli* produce lactic acid (by fermenting lactose [milk sugar] in the milk) as a coagulating agent (which makes the yogurt thick) and preservative, but also grow in incredibly large numbers (100,000,000 bacteria/g) in tandem with the *Streptococci* to outgrow any potentially dangerous impostors. Yogurt production was originally used, perhaps first in Turkey, as a means of storing milk for the winter months when "traditional" cows, goats and sheep don't produce milk (Fig. 3-3).

Pathogens

There are 7 main groups of pathogenic bacteria important in food safety. I will deal with them in turn.

Fig. 3-4. An electron micrograph of *Campylobacter* [magnification approx. 25,000Å~] showing its spiral shape and flagella (kindly provided by Manfred Ingerfeld, University of Canterbury, New Zealand)

Campylobacter

This is a Genus of spiral-shaped bacteria with flagellae at both ends. They are very mobile, by flapping their flagella, and swim over wet surfaces spreading quickly (Fig. 3-4). Most human disease is caused by *C. jejuni*. Campylobacter lives and reproduces best in birds (they like the bird's body temperature); birds carry the bacteria without becoming ill. Surprisingly it is a very fragile bacterium – I find it hard to understand how it can be such an enormous problem in food – requiring very precise growth conditions. It is killed by oxygen, cannot tolerate drying, and is killed by freezing. It proved very difficult indeed to culture in the laboratory until special culture media and conditions were devised by Professor Eric Bolton and his colleagues in England.

Campylobacteriosis

This is the disease caused by *Campylobacter* infection. Most people who get campylobacteriosis have diarrhoea, stomach cramps and fever. The diarrhoea is sometimes bloody and accompanied nausea and vomiting. It takes 2–5 days after exposure to the bacterium to develop the disease. So if you get campylobacteriosis it cannot have been caused by last night's dinner from the takeaway – you will have to think about what you ate at least 2 days ago to try to identify the cause. Campylobacteriosis usually lasts for a week, most people recover completely, but a very few people with an impaired immune system can develop blood infections and die. The disease is notifiable (i.e. must be reported to the public health authorities) in many countries.

Campylobacteriosis is one of the most common bacterial causes of diarrhoea. In the USA its incidence is 15 cases/100,000 population/year (i.e. 0.015% of Americans will suffer from the disease each year). In other countries it is more common. For example in developing countries it is very much more common, and surprisingly New Zealand is at the top of the developed worlds campylobacteriosis league, and frighteningly close to countries with less well developed public health systems (Fig. 3-5).

There is usually no treatment for campylobacteriosis because sufferers generally recover quickly. However if it is caught early enough antibiotics can reduce the duration of the disease.

Which Foods Does *Campylobacter* Come from?

There is a great deal of scientific research underway throughout the world to find out where this virulent food poisoning bacterium comes from. Chickens have received a very great deal of attention because it is well known that *Campylobacter* infects chicken flocks without causing the birds any ill effects. It resides in their guts, and when they are slaughtered might contaminate the flesh. If the chicken is not cooked properly the consumer might get infected. Thorough cooking will definitely kill the bacterium – remember it is a fragile organism with very exacting requirements to survive. Work in New Zealand has suggested that using tongs to barbecue chicken might transfer *Campylobacter* from the uncooked to the cooked meat when the same

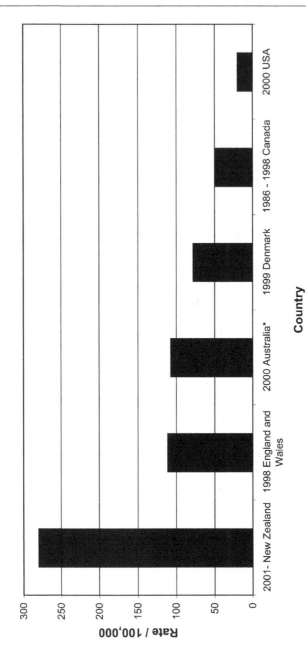

Fig. 3-5. Campylobacteriosis incidence around the world (data kindly provided by Dr Andrew Hudson, ESR, New Zealand)

tongs are used (as they usually are) to handle the chicken throughout the barbecuing process. This is only one possible transmission route, but it illustrates the way in which the disease might be transferred by simple cooking procedures. The poultry industry thinks that we are being unfair to single their product out for blame. They are in part justified in this opinion because *Campylobacter* can also be found in water, pet faeces, and even milk ... so there are many potential sources. However campylobacteriosis is often associated with chicken consumption. In fact in Belgium during the recent dioxin scare (I'll discuss this later – see Chapter 9) there was a concern that dioxin-contaminated chicken feed might have led to unacceptable levels of dioxins in chicken meat, and so chicken was withdrawn from sale for a few weeks. During this time the incidence of campylobacteriosis went down. I find this compelling evidence that chicken and campylobacteriosis are linked. But to be certain we must wait for the huge global research effort to explore all of the possibilities and report on their findings. In the meantime cook your chicken well and be careful which tools you handle it with and you shouldn't be one of the 0.015% of the population lying in bed with a campylobacteriosis tummy ache regretting that you went to your best mate's barbecue!

Salmonella

Edwina Curry's enthusiasm (see Chapter 2) for *Salmonella* has made this group of bacteria a household name. This is a large genus of bacteria, two are important as food poisoning agents, *Salmonella typhimurium* and *S. enteritidus*, and others are very important human pathogens generally not necessarily associated with food, for example *S. typhi* causes typhoid fever – water is usually the transmission route. They were discovered by a Dr Salmon (hence their name) in the USA about 100 years ago.

The *Salmonellae* are rod-shaped bacteria that prefer to live without oxygen (facultative anaerobes). They naturally inhabit the intestine of humans and animals. Many species are harmless, indeed they have an important role as part of the gut's complex bacterial ecology which is part of the food digestive process.

Salmonellosis

The symptoms of diarrhoea, fever and stomach cramps appear 12–72 hours after eating contaminated food. Most people recover completely within 7 days, however a small proportion of infections lead to extreme diarrhoea which requires hospitalisation and intra venous fluid administration. In the latter group it is possible that the *Salmonella* might move from the gut into the blood stream so causing a potentially fatal septicaemia unless antibiotic treatment is given quickly.

In the USA about 40,000 cases of salmonellosis are reported each year (i.e. 0.013% of the USA population will get the disease each year). This is almost certainly a gross underestimate (the real figure might be 30 times larger) because mild disease is likely not to be reported – sufferers don't feel ill enough to go to their doctor and so don't become a statistic.

Which Foods Does *Salmonella* Come from?

Since *Salmonellae* are common bacteria of the guts of many animals, including humans, they are easily transferred to meat in the slaughter house, and by cooks and food handlers who do not practice good personal hygiene. However cooking very effectively kills these bacteria and so contamination should not present a food safety issue providing the contaminated food is cooked properly. *Salmonella* is usually associated with meat, but vegetables might also be contaminated following the use of animal fertilisers or contamination with soil that might also contain *Salmonellae* (presumably originally from animal faeces).

There has been a great deal of publicity about *Salmonella* in eggs. *Salmonella* can contaminate eggs in two ways. The commonest is for the bacteria to be on the outside of the shell – eggs are laid via the chicken's cloaca (an anatomical chamber in birds into which faeces and urine pass before being expelled) and therefore are likely to get a coating of *Salmonella* from faeces. When the egg is cracked before being used in cooking it is possible that the yolk or white might get contaminated. *Salmonellae* can also be incorporated into the egg via infection of the bird's oviduct. In this case the inside (i.e. yolk and white) of the egg are contaminated. Which ever way the egg is contaminated if it is used in recipes (e.g. mayonnaise) that do not involve sufficiently high temperatures to kill the bacteria, the product poses a significant risk to the consumer, especially if the food is not consumed straight away and therefore the bacteria can multiply.

Salmonella can also be transferred from food to food. For example, if you cut some meat up on your chopping board, forget to wash the board thoroughly, then cut some boiled ham for a sandwich to eat for your lunch at work, the *Salmonella* might cross infect the meat, grow in your sandwich in its nice warm, humid lunch box, then find the perfect conditions to multiply in your intestine....a day or two later you're in bed with a zillion *Salmonellae* in your gut making their presence known.

E. coli

As discussed previously *Escherichia coli* is a species of bacteria that has many different strains. Most are harmless, normal components of the complex mixture of micro-organisms that inhabit our gut. However, several strains are pathogenic – these are the Shiga-toxin producing *E. coli* (STEC). The most common STEC in the USA is *E. coli* O157:H7 (the "O157:H7" denotes the specific pathogenic strain – the numbers and letters refer to specific cell surface markers found on the *E. coli* cell). This is an emerging food born illness that is slowly moving around the world.

The most important difference between STECs and 'normal' *E. coli* is the production of Shiga toxins by STECs. Shiga toxin is a bacterial toxin produced by a genus called *Shigella*. Bacteria are able to mate (conjugate) during which genes pass between the conjugating bugs. At some stage an *E. coli* might have conjugated with a *Shigella* and the latter passed its Shiga toxin gene into the *E. coli* thus making the latter pathogenic due to its acquired ability to secrete this highly toxic chemical. There is also the possibility that bacterial viruses (bacteriophages) that cross infect *Shigella* and *E. coli* transferred the Shiga toxin gene. This is exciting science about which microbiologists are still uncertain, but very actively researching.

Infection with *E. coli* O157:H7

The usual symptoms are bloody diarrhoea and stomach cramps. In rare cases kidney involvement can result in a severe, often fatal, form of the infection. This happens because red blood cells are destroyed by the infection which results in kidney failure – this is called haemolytic uremic syndrome. In the USA about 2–7% of cases result in kidney failure, indeed this is the principal cause of kidney failure in children.

About 73,000 cases and 61 deaths occur annually in the USA (i.e. 0.024% of the population get STEC infection/year).

Which Food Does *E. coli* O157:H7 Come from?

In the USA, where the first cases were identified in 1982, the main source of infection was ground beef (e.g. used to make hamburgers). Providing the meat is cooked thoroughly there is no problem because the bacterium is killed by heat, and the toxin is inactivated. However many Americans like their hamburgers pink in the middle, and this is where the *E. coli* O157:H7 are lurking because the temperature does not get high enough to kill them. Ground meat is a particular problem because *E. coli* O157:H7 occurs only on the outside of meat – it gets there as a faecal contaminant in the slaughter house. If a steak is contaminated and well cooked on the outside the surface bacteria will be killed, even if the inside is pink. However, when the beef is ground the bacteria are mixed into the mince and therefore only cooking well on the outside is not good enough to kill them.

Any food that might come into contact with animals' faeces can be contaminated with *E. coli* O157:H7. Milk is a good example – this is another good reason to drink pasteurised milk. There was a case of a child getting serious *E. coli* O157:H7 in New Zealand recently. The child drank un-pasteurised milk from the family cat's bowl. There are several possible transmission routes here:

1. The un-pasteurised milk might have been contaminated with *E. coli* O157:H7 from the cow.
2. The *E. coli* O157:H7 might have originated from the cat – if you have a cat, watch it groom....the potential for faecal bacteria to contaminate its mouth will become obvious!

As mentioned above, this is an emerging disease its incidence around the world is increasing. It will get worse. We will hear a lot more about it over the next few years.

Clostridium

Clostridia are rod-shaped oxygen hating (anaerobic) bacteria that are able to produce spores. These spores are incredibly resistant – they will survive boiling – and represent a dormant phase that can germinate many months, or perhaps years, later to liberate the active bacterium. Spore formation is a means of protecting an organism against adverse conditions, so *Clostridia* will sporulate (form spores) if their living conditions are not to their liking. For example, if the temperature is too high, or the medium to dry.

If *Clostridia* infect food they will grow if the conditions are right, but if they don't like the environment (e.g. the food is in the fridge) they will form spores. If the food is taken from the fridge and cooked, the spores might survive and germinate later when the cooked food is stored in conditions that the bacteria like. This is a big problem because it means that even cooking does not kill the bugs.

If being able to form spores is not bad enough, these creatures also produce poisons of amazing toxicity (I'll discuss their toxicity later). It is the toxins that cause the diseases associated with eating food contaminated with *Clostridia*. They can cause problems in two ways. If you eat food contaminated with the active bacteria, they can grow in your gut (the conditions are ideal, 37°C, humid, and oxygen free) and produce toxins. On the other hand if you eat food containing toxins because it once was infested with active bacteria, if the toxin concentration is high enough you will become ill. So you don't need to eat the bacterium itself to get ill. *C. botulinum* can cause disease by both gut infection and toxin ingestion, but *C. perfringens* almost always acts by the gut infection route.

There are several *Clostridia* important in food safety. One, *Clostridium botulinum*, is arguable the most dangerous bacterium. *C. botulinum* caused botulism, an often fatal food born disease, and *C. perfringens*, a relatively common food bacterium that is rarely fatal.

Clostridium Botulinum

Disease caused by *C. botulinum* (botulism) is very serious indeed, it is considered a medical emergency. The *C. botulinum* toxin (botulinum toxin) is an incredibly potent nerve poison, its lethal dose in people is about 1 millionth of a gram [µg] – you would need a microscope to see this amount. It is 5,000 times more toxic than potassium cyanide. Infection with *C. botulinum* or ingestion of botulinum toxin results in respiratory failure that requires the patient to be given assistance with breathing while their body eliminates the poison. If medical support is not given death is inevitable. About 8% of infected people die in countries where advanced medical support can be given quickly.

Botulism

Botulism is not only caused by food, it can also result from wounds infected with *C. botulinum* and consumption of *C. botulinum* spores from non-food sources (this is commonest –

but still rare – in children who eat all sorts of unspeakable things). In the USA there are about 110 cases of botulism each year, 25% of these are foodborne (i.e 0.000009% of the population get foodborne botulism each year – it is a very rare disease).

Most cases of botulism in the States are due to home canning (bottling) of food. All that is needed is a tiny contamination of soil with *C. botulinum* bacteria or spores and these will grow very nicely in the nutrient rich preserved food on the pantry shelf. Such cases are sporadic. There are occasionally outbreaks of botulism caused by commercial food contamination.

A botulism outbreak occurred in the UK in 1989. It involved contaminated hazel nut yogurt; the hazel nut puré used in the manufacture of the yogurt was contaminated with *C. botulinum*, possibly because the nuts were collected off the ground and were contaminated with soil. Twenty-seven people were seriously ill, and one died. This illustrates the serious problem that contamination of mass-produced food can cause.

Botox

A discussion of botulinum toxin is not complete without mention of wrinkles and vanity. The connection between serious, life threatening food poisoning and wrinkles is perhaps a little obscure, but the much heralded wonder drug of some vain oldies is indeed botulinum toxin under a new name – Botox. It is a nerve poison and if injected into muscles causes them to relax (because it inhibits nerve impulses to the muscles). If you inject the stuff into wrinkly skin the effect on the underlying muscles results in magical disappearance of the wrinkles. Usually therapists focus on their clients' faces, the Botox gives their beautified client a non-wrinkly but frighteningly expressionless appearance. I think I'd rather be wrinkly! The dose of botulinum toxin used in Botox therapy is infinitesimally small, but I have always wondered what would happen if it was accidentally injected into a blood vessel... I'm sure someone has thought of this though – or at least I hope they have.

Clostridium Perfringens

Perfringens food poisoning is unpleasant, but not serious. It is characterised by stomach cramps and diarrhoea that begin 8–22 hours after eating infected food. A day in bed is usually sufficient for a return to normal health. As with *C. botulinum*, *C. perfringens* produces a toxin that causes the illness. Perfringens toxin is very, very much less potent than botulinum toxin.

There is a more serious, but rare, form of perfringens food poisoning that is caused by a different strain of the bacterium – *C. perfringens* Type C. It caused enteritus nercoticans or pig-bel which involves bacterial-induced death of the cells of the intestine. This is extremely serious and is usually fatal.

Perfringens food poisoning is common in foodborne illness terms. There were 1,162 (i.e. 0.004% of the population get the disease each year) cases in the USA in 1981. This reported incidence rate is probably very much lower than the real incidence because the disease is not serious enough for most people to go to their doctor. In addition, most doctors would not take faecal samples to confirm the causative bacterium. The Centre for Disease Control (CDC) in the States estimated that the real incidence rate is likely to be closer to 10,000 (i.e. 0.04% of the population) cases per year.

So what are the likely sources of this kind of food poisoning? Basically any food that is prepared hours before serving is the danger zone for perfingens. An example might be a sauce made from meat juices. The meat juices could contain *C. perfringens* spores that have survived cooking, the spores then sit in the sauce in a nice warm kitchen, germinate and the released bacteria grow. In the right conditions a bacterium can divide every 20 minutes, so if you start with 500 spores in the sauce, and the sauce sits in the kitchen for 24 hours, there could be 36,000 toxin-producing bacteria ready to colonise the consumer's gut where they will continue to produce toxin and result in perfringens food poisoning. Institutional (such as schools, nursing homes, prisons, etc.) food is most commonly associated with this kind of food poisoning.

A large outbreak of perfringens food poisoning occurred in the USA in 1985. It involved a celebratory banquet in a factory in Connecticut. The meal involved gravy that had been made the day before and only warmed up before serving. Of the 1,362 people at the banquet, 599 (44%) succumbed to *C. perfringens*. The Chef learned a very hard lesson about food safety – never warm up a sauce, bring to the boil and keep it at a high temperature for a few minutes. This will kill the active *C. perfringens* that might be lurking, just waiting to get into your gut and excrete toxin.

Listeria

There is only one species in this genus that is important as a foodborne illness – *Listeria monocytogenes*. It is a rod-shaped bacterium with flagellae that make it motile. It likes to live and grow at cool temperatures which is a problem because refrigeration of contaminated food can give it exactly the conditions that it likes best. Most people think that refrigeration of food is a good way to reduce food poisoning, this is generally so, but not for food infected with *L. monocytogenes* which grows well at refrigerator temperatures (+4°C) (Fig. 6).

Listeriosis

Infection with *L. monocytogenes* causes listeriosis. This is a serious disease that can be fatal in children, people whose immune systems are not up to par; it is particularly dangerous in pregnancy when it might cause abortion or birth defects. The symptoms of infection are fever, muscle aches, and sometimes nausea and diarrhoea. However, the bacterium can get into the central nervous system (CNS – brain and spinal cord) the resulting disease is very serious being characterised by headaches, stiff neck, confusion, loss of balance, and in severe cases convulsions.

About 2,500 people a year get listeriosis in the USA (i.e. 0.0008% of the population), of these 500 (i.e. 20%) die. This is a serious, but uncommon, disease. Twenty percent of listeriosis

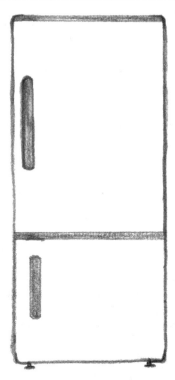

Fig. 3-6. *Listeria* grows best at cool temperatures, so the fridge is the place it likes to be

occurs in pregnant women. So they are both more susceptible to the disease and at greater risk from it. Their increased susceptibility is probably because their immune systems are challenged by pregnancy. The challenged immune system hypothesis is further supported by the fact that AIDS sufferers are 300 times more likely to get listeriosis than any one else. Healthy adults occasionally get infected, but they do not usually get serious listeriosis and therefore recover quickly.

Where Does *Listeria* Come from, and How Does it Get into Food?

As with so many other food bacteria *L. monocytogenes* is a soil bacterium, probably getting there from animal faeces. It can live happily in the guts of animals (e.g. cows) without causing them any apparent harm. It can get into food either via soil contamination (e.g. of vegetables that are eaten raw), or via faecal contamination of meat at the abattoir, or of milk during milking (this is why hygiene in the milking parlour is so important).

Listeria is killed by pasteurisation – this is one of the arguments for pasteurised dairy products, including cheeses. There have been problems with some of the wonderful European soft cheeses (e.g. Brie) being *Listeria* infected, at one stage the UK press coined the phrase "*Listeria* hysteria" to describe the panic over the possible ill effects of eating cheese, especially for pregnant women. Still, pregnant women are advised to steer clear of soft cheeses – quite rightly; this is a good risk risk-reduction strategy.

So providing you only eat cooked or pasteurised foods, particularly dairy product, you won't get listeriosis – WRONG! As I mentioned above *Listeria* grows well in the fridge, for this reason it is a great problem for food manufacturers and food retailers. For example, if a cold cooked meat manufacturer cross contaminates *Listeria*-infected uncooked pork and sterile boiled ham, then slices the ham for packaging, it is possible that hundreds of packets of sliced boiled ham will be *Listeria* infected. The packed ham will then be sent to shops and wait in the chill cabinet for some unsuspecting shopper to buy one for their sandwiches. The *Listeria* is likely to multiply like crazy on the chilled ham and the poor ham sandwich eater will get a few million *L. monocytogenes* with their lunch, and might get listeriosis. Listeria is a manufacturers nightmare, because the conditions that are normally good practice (i.e. low temperatures) are *Listeria* heaven.

There are many examples of *Listeria* outbreaks resulting from contamination by manufacturers. For example, there was an outbreak in the USA in 2002 that resulted in 46 cases, 7 deaths, and 3 stillbirths/miscarriages spread over 8 States. It was traced back to a single source of cold cut turkey. As we move further and further towards large scale manufacturing this type of incident is likely to get more common, unless we tighten up hygiene practices – for example, post-packaging pasteurisation.

Listeriosis seems to have peaked, the number of cases in the USA is in decline. This is because we know more about how the bacterium spreads and how to reduce food contamination.

There has been a decline in cases of listeriosis in the USA (*data from CDC, Atlanta, http://www.cdc.gov/ncidod/dbmd/dis-easeinfo/listeriosis_g.htm*)

- 1989–1993 44% decline
- 1996–2001 35% decline.

Staphylococcus

Staphylococcus is a genus of spherical bacteria (cocci – from the Greek *kokkos* meaning berry) that usually grow in pairs, short chains, or bunches (a bit like grapes). Some produce a very heat stable toxin.

Staph Food Poisoning

Staph food poisoning (staphyloenterotoxicosis or staphyloen-terotoxaemia) is caused by toxin-producing strains of *Staphylococcus aureus*. The bacterium itself does not do much damage, it is the toxin that causes the disease – the non-toxin forming strains of *S. aureus* are not usually troublesome pathogens. The disease is usually caused by eating food contaminated with the toxin due to growth of *S. aureus* on the food, for this reason the time between eating the contaminated food and onset of the symptoms is very short. The disease is ostensibly poisoning not microbiological.

Ingestion of enough of the toxin (i.e. 1 µg – a millionth of a gram; staph toxin is incredibly potent) causes acute nausea, vomiting, retching, stomach cramps, and the desire to lie out flat (prostration) to reduce the stomach pain. Although onset is rapid, recovery can take 3 or more days according to the dose of the toxin. A dose of 1 µg of staph toxin equates to food with 100,000 *S. aureus*/g – this is a fairly high bacterial contamination level. Death from staph food poisoning is very rare.

It is difficult to be certain how many cases of staph food poisoning occur each year, because many people do not go to their doctor, and because the symptoms are similar to those of other food poisoning bacteria. Only a test for the toxin can identify the disease with certainty. Despite this the USA had 1,257 cases in 1983 – these were associated with just 14 outbreaks (i.e. sources of contamination). There has been a decline since 1983 with only 100 cases confirmed in 1987.

Which Foods Are Associated with Staph Food Poisoning?

Foods that are handled a lot in their preparation and that might be kept at room temperature for a long time before being eaten are the highest risk. These include:

- Meat/meat products
- Poultry and egg products
- Egg, tuna, chicken and potato salads
- Macaroni
- Cream-filled pastries, cream pies, chocolate éclairs
- Sandwiches
- Dairy products

Where do Staphylococci Come from?

They are natural flora of our noses and throats, and are commonly found on hair and skin. They get into food by unhygienic handling practices – touching food without properly washing your hands; coughing on food; hairs dropping into food. Once the food is contaminated the bacteria grow and produce toxin during its storage. It does not take long for a few bacteria to get to the 100,000/g threshold that can cause illness.

With a doubling time of 20 min, a hair with 1,000 S. aureus on it dropped into a tuna mayonnaise sandwich would need only 2.5 hours to produce food poisoning levels of toxin in the sandwich.

In the USA 1,364 out of 5,824 (23%) kids from 16 elementary schools in Texas became ill soon after eating their school lunch. The lunches were prepared at a central kitchen and transported to the schools in the region. Most of the children who became ill had eaten the chicken salad lunch choice. It is clear from a detailed analysis of the preparation technique why the kids got ill. The onset of symptoms was rapid which pointed to staph food poisoning; the suspicion was confirmed by finding staph toxin in the chicken salad.

So what happened during the preparation process that resulted in 1,364 kids getting staph food poisoning? The most likely scenario is that the chicken was contaminated during deboning, it was then cooled too slowly which allowed the contaminating *S. aureus* to grow and make toxin. The bacteria had another chance to grow when the food was kept in the classrooms waiting for lunchtime.

A HACCP analysis (see Chapter 1) might have prevented the outbreak. The critical control points are cooling (1) and storage (2) – cool quickly and store cooled. This might have prevented the incident. But where did the contamination come from in the first place? I wonder whether the cooks wore gloves, or washed their hands properly? We can never be certain (see also the table).

	Sequence of events	Critical control points
Day 1	Frozen chickens boiled for 3 hours	
	Chicken cooled with a fan, taken off the bone, ground	
	Refrigerated (+4 °C) overnight	1
Day 2	Chicken mixed with other ingredients	
	Transported to schools in cool, insulated containers 9:30–10:30am	
	Stored at schools at room temperature	2
	Eaten at 11:30am–12:00noon	

Case report from The Bad Bug Book, Chapter 3, US Food & Drug Administratio (2003)

Bacillus

The most important food poisoning *Bacillus* is *B. cereus*. It is a spore-forming rod-shaped bacterium that prefers low oxygen conditions, but can grow in the presence of oxygen (facultative anaerobe).

Bacillus Emetic and Diarrhoeal Food Poisoning

B. cereus can produce two toxins. One is a large heat-labile protein that causes diarrhoea, the other is a small heat-stable protein (peptide) that causes vomiting. Because of the two toxins, there are two distinct diseases caused by *B. cereus*; diarrhoeal, and emetic food poisoning. The former has symptoms just like *C. perfringens* food poisoning with an onset 6–15 hours after ingestion of infected food. The emetic type is much quicker, symptoms appear 0.5–6 hours after eating a contaminated meal, its symptoms are like staph food poisoning. Both diseases clear up in about 24 hours and are not particularly severe.

Which Foods Does *B. cereus* Come from?

The diarrhoeal disease is associated with many different foods, but the emetic disease is almost exclusively associated with rice and shellfish. Although recent cases in New Zealand resulted from the consumption of potato topped pies made from reconstituted dried potatoes.

The common feature of *Bacillus* emetic food poisoning is alkaline carbohydrate foods that have been cooked, stored and reheated. *B. cereus* does not grow well in acid conditions, so all that you need to do to minimise the risk of the bacterium growing is to make the food acid – sprinkle a few drops of lemon juice onto your cooked rice if you want to store it to make fried rice the next day. The reason for these food types and conditions being just right for *B. cereus* is that, for example, if you cook rice and the rice gets contaminated, the bacterium will grow during the cooling and storage period and produce its heat stable pep-

Fig. 3-7. Re-cooked rice is a good place to find *B. cereus* (*from www.eeecooks. com/ recipes/2002/02/06/fried_rice_1728.jpg, by kind permission*)

tide toxin. When you use the rice to make fried rice or kedgeree the bacteria are killed but the toxin is not destroyed. When you eat your re-cooked rice you get a dose of toxin, and within no time at all you don't feel too well! (Fig. 3-7).

Other Food Pathogens

I have discussed only the most commonly encountered bacteria that cause food poisoning. There are many more and many microbiologists would take issue that I have left out an important pathogen. So for completeness, here are some of the other genera that can cause problems in food: *Yersinia, Vibrio, Aeromonas, Pleisiomonas, Shigella* and *Streptococcus*… and I'm sure there are more, many of which have not yet come to light. Remember that bacteria are constantly exchanging genes to form new strains (serotypes), there are so many awful combinations of bacterial virulence and toxin production that could create the most horrific food pathogen … it's just a matter of time!

Fig. 3-8. New Zealand's cook, cool, chill campaign – aimed at making people think about hygiene in their own kitchens (reproduced by kind permission of the New Zealand Food Safety Authority)

4
Viruses

4 Viruses

Viruses are the smallest living things. There is serious scientific discussion about whether they are just a collection of chemicals or whether they are actually living. Many scientists now regard them as life, but in its simplest, most economical form. Usually living organisms have a myriad enzymes, structural proteins, lipids, complex sub-cellular structures and almost unbelievably complex biochemical processes to provide the energy to allow growth, development and reproduction. Viruses short cut all of this by living inside host cells and utilising their biochemical processes to survive and reproduce. The primary purpose of a virus is to infect a cell and to reproduce. A successful virus will not kill its host cell, or at least not until after it has successfully reproduced, and released more viruses to infect more cells and start the process over again.

The Discovery of Viruses

Pasteur's work on bacteria showed that fine filters would retain bacteria, so it was possible to filter out infectivity – this is still used today as a means of sterilising liquids using very fine (e.g. Millipore®) filters. In 1882 Dmitry Ivanowski (1864–1920), a Russian microbiologist, studied tobacco mosaic disease and showed that the agent that caused the disease could not be filtered out. Indeed, he showed that cell-free filtrates could cause the disease in healthy tobacco plants. When he tried to culture this disease causing principle in culture media used for bacteria, he found that no growth occurred. So he had isolated a very small (i.e. it would pass though a filter that retained bacteria and other cells) disease causing agent that could not be cultured. A prominent Dutch scientist, Martinus Beijerinck (1851–1931), confirmed these findings and called the infectious filtrate "living infectious fluid". Pasteur was investigating rabies at about the same time (1892) and called the filterable agent that caused rabies "virus". It was about 25 years before instrumentation became sufficiently advanced to allow proper study of the miniscule disease causing "viruses" that Pasteur, Ivanowski and Beijerinck had described. By this time it was known that viruses could not be grown outside living cells. It is for this reason that most of the work on the structure and infectivity of viruses was done on bacteriophages because they were easy to culture in bacteria. At this point in scientific history it was not possible to culture mammalian cells and therefore human and animal viruses could not be cultured outside living creatures.

It was not until the 1960s when the electron microscope was introduced, and animal cell culture techniques became available that the first images of viruses were seen. They were a marvel; beautiful, symmetrical, structures containing the rudimentary principles of life. But were they alive, or were they just a collection of chemicals?

Some of the Worst Diseases are Caused by Viruses

Viruses can be very nasty indeed. They cause some of our most feared diseases. Smallpox, Ebola, Acquired Immune Deficiency Syndrome (AIDS), Severe Acute Respiratory Syndrome (SARS), and polio are all viral diseases. Some cancers are caused by viruses – I'll talk about these later. Worst still, no viral disease is curable by medicines, although there are now pharmaceuticals available that slow down, or inhibit, viral reproduction so slowing down the course of the disease that the particular virus causes. An example of such a medicine is Acyclovir used in the treatment of AIDS and some hepatitis virus infections. Usually, as with the common cold, you just have to wait until the body's own defences (the immune system) overpowers the virus. Contrary to popular belief, antibiotics have no effect on viruses. Viruses are the ultimate infectious agents, they can't be treated, and some trick our immune system into not recognising them, or interfere with the process of immunity in order to allow themselves to reproduce without being wiped out by antibodies produced by their host. AIDS works in this way. It suppresses the immune system which is why it is so devastating.

The picture of viruses that I have painted so far is particularly grim, but there are some good viruses. These are a group of very specifically adapted viruses that attack and kill bacteria. They are called bacteriophages, or phage for short. They are becoming increasingly important as a potential means of destroying pathogenic bacteria, for example on food. I'll return to this later too.

What is a Virus? How Does it Infect a Cell?

There are several types of virus, but basically they all consist of a lipoprotein coat with nucleic acid inside. They are classified according to the type of nucleic acid that they contain. So there are DNA (deoxyribosenucleic acid) viruses and RNA (ribosenucleic acid) viruses. They are incredibly small; most are between 20 and 400 nm (a nanometre is 0.000000001; 1/1,000,000,000; 10^{-9}, or a thousand millionth of a metre). This is the size of a large molecule, and therefore very powerful electron microscopes are needed to see them (Fig. 4-1).

To infect a cell the virus lipoprotein coat fuses with the cell membrane (a protein lipid "wall" that holds the cell together) releasing its nucleic acid (DNA or RNA) into the cell. The nucleic acid then incorporates into the cells own genome (i.e. collection of genes also made from DNA) and gets replicated when the cell switches on its own DNA replication. The viral nucleic acid codes for its own lipoprotein coat, so that as the host cell replicates DNA and makes proteins from the DNA template it produces all that the virus needs to make more viruses. When

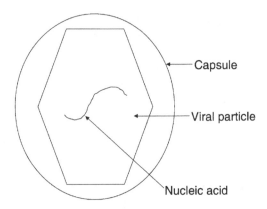

Fig. 4-1. The structure of a typical virus

the virus has replicated the host cell holds many thousands of viral offspring. At this point the cell dies, ruptures and releases the viruses to infect surrounding cells, or be expelled from the body and infect a new unsuspecting host. This is what happens when you have a cold, you sneeze out viruses that are breathed in by someone else and grow in the new host's cells.

Viruses and Food

Viruses are gaining increasing importance as food borne vectors of disease. It is estimated that in the USA at least 75% of food poisonings are caused by viruses. Many viruses cause rapid onset, short, sharp illness, and so by the time a sufferer gets round to going to their doctor the symptoms have subsided. For this reason the vast majority of food borne viral disease is not reported, and therefore it is not possible to estimate its real incidence.

There are several viruses important in food safety. I'll look at them individually.

Noro Virus

This used to be called Norwalk-like virus – it was re-named in 2002; Norwalk is a town in Ohio, USA where the first outbreak occurred in 1968. It is a member of a family of viruses called the *Caliciviruses* which have their genetic material on single stranded DNA.

As we learn more about the virus it is becoming clear that it is probably the most prevalent food poisoning organism world-wide. In the USA there are at least 180,000 confirmed cases per year, and it is estimated that 60% of the US population is exposed to the virus by the age of 50. It has been estimated that in the USA there are 9,200,000 Noro virus cases per year (i.e 3,150/100,000 population/yr, or an incidence of 3% disease in the population). This makes Noro virus by far the commonest form of food poisoning.

There are a range of Noro viruses, each is slightly different (different serotypes, i.e. they cause different antibodies to be formed following inoculation of animals/humans), but they all cause pretty well the same symptoms. Typical symptoms of Noro infection are vomiting and watery diarrhoea, sometimes with stomach cramps and nausea, and less frequently with fever. The virus acts quickly – symptoms begin 24–48 hours after infection, and have usually cleared up within a further 24 hours. Death has been reported in rare cases where the sufferer was unable to replace the fluids lost quickly enough.

Noro viruses can only live and reproduce in human cells. The virus is present in the faeces of infected people, therefore the faecal/oral route is the most important transmission route. This is the typical food borne illness transmission route – an infected person goes to the toilet, does not wash their hands properly (if at all) then handles food. The consumer of the food is very likely indeed to get infected. The worst scenario is a person who is infected, but has not started to show the symptoms and therefore does not know that they present a particular hazard if they work in the food industry.

There are numerous examples of Noro virus outbreaks. Most are in institutions, cruise ships, or in hotels. There are "captive audiences" here, all ready and waiting to be infected! All that it takes is one infected person to handle food and the virus will spread through the "closed" community very quickly indeed. Another faecal/oral infection route is via shellfish that might be growing in infected waters.

Shellfish

New Zealand had a particular problem with this route in 2001/02 when mussels and oysters grown in Northland were contaminated because pleasure boats moored near to the shellfish farms were discharging untreated sewage, Noro virus in the sewage contaminated the sheltered harbour, and the virus was filtered by the shellfish. Anyone eating the uncooked shellfish was at significant risk of infection. The contamination rate was so high that the shellfish farms were closed until the problem could be solved. They were still closed in 2004.

A specific case illustrates the problem well. Thirty-six of 95 people (38%) attending a yacht club Christmas celebration in the north of North Island, New Zealand in December 1994 went down with gastrointestinal illness likely to be Noro virus. The oysters were very likely to have originated from the Bay of Islands in Northland, New Zealand. There are numerous similar epidemiological examples that point to oysters and the Bay of Islands, hence the draconian, but justified, action to close the fisheries (see also page 91).

Cruise Ships

An outbreak of Noro virus infection occurred on the Star Princess which left Seattle in August 2003. Within the first 2 days of the cruise 60 of the ship's 2,800 passengers (2%) reported gastrointestinal upset. Clearly something was wrong! Simple hygiene regulations were activated which resulted in prevention of further spread of the virus. The source of the infection is not known, but it is most likely that one of the kitchen staff was infected, or less likely that the infection originated in a passenger who handled food that others ate.

Restaurants

There have been numerous cases of Noro virus infection originating from an infected staff member in fast food outlets. Less commonly diners are implicated, but there is one very good example of this, it occurred in Derby, UK in a hotel dining room. A diner was suffering from Noro virus infection and vomited in the restaurant. Within 48 hours over 60% of the restaurant's diners were suffering from gastroenteritis. The most likely transmission route here is virus-containing droplets originating from the vomit landing on the diners meals. This is supported by the fact that the diners nearer to the woman were more likely to get ill; 90% of the people at the same table were infected.

Retirement Homes

This is another captive population, and there are numerous cases of Noro virus infection sweeping through rest and retirement homes. The problem here is that the residents are likely to be severely affected; they are old, less likely to produce a good immune response, and more likely to suffer the ill-effects of dehydration. These are the occasions that are most frequently associated with fatalities.

How to Prevent Infection

There are a set of simple rules (formulated by the Virginia Department of Health, USA) that food handlers, and others should follow to reduce the risk of the spread of Noro virus infection:

- Wash your hands frequently
- Promptly disinfect contaminated surfaces with household chlorine bleach based cleaners
- Wash soiled clothing
- Cook oysters completely to kill the virus
- Avoid food (or water) from sources that may be contaminated. _____

Hepatitis Viruses

Hepatitis (from the Greek, *hepatikos* – liver) is inflammation of the liver. It can be caused by chemicals (e.g. alcohol), bacteria or viruses. Only viral hepatitis will be discussed here.

There are 5 different hepatitis viruses, they are called Hepatitis A, B, C, D and E. Only HepA can be food borne, but this is not its only means of transmission.

HepA

The HepA virus (HAV) is a member of the family of viruses called the Picornaviridae, they have a single RNA molecule surrounded by protein capsule (called a capsid) with a diameter of 27 nm. Most of the Picornaviridae cause disease, others include the polioviruses and rhinoviruses that cause the common cold.

HepA is not a serious disease – unlike the other types of hepatitis which are often very serious, and might have long term effects like cancer or cirrhosis. Some might be fatal. The symptoms of HepA include sudden onset of fever, malaise, nausea, anorexia, and abdominal discomfort, followed after a few days by jaundice (yellowing of the skin and whites of the eyes). The disease usually lasts for less than 2-weeks with complete recovery. Only 10–100 viral particles are needed to cause infection.

HAV lives and reproduces in the liver and is secreted into the bile which is released into the intestine via the bile duct. The virus can survive in the intestine and passes through with the food, eventually being expelled in the faeces. For this reason faeces from HepA sufferers are infectious.

Route of Infection

HepA is primarily a food borne disease. A HepA sufferer who is not hygienic after going to the toilet and then handles food is likely to leave viruses behind on the food. This is the same story as for Noro virus. Don't forget that it takes less than 100 virus particles to cause an infection. Faeces from an infected person would harbour many millions of viruses.

HAV appears to be more stable than many other viruses. Most viruses don't survive long outside a host cell, however HAV can survive for days, so contaminated food remains infectious for a few days, if not longer. Cooking kills the virus, so transmission usually involves foods that are not cooked (e.g. fruit picked by infected people), or cold cuts of meat.

HepA in Blueberries

New Zealand's North Island is a producer of excellent blueberries. At Christmas 2001 there were several cases of HepA in New Zealanders. This was not considered to be of any great importance initially because we would expect a few cases of HepA in the population each year, and the disease is not serious. However as the number of cases rose to 29, the Health Authorities became more interest because of the possibility of a point source of contamination, and the need to eliminate the source in order to prevent further spread of the infection.

A great deal of good epidemiological work led to identification of a common factor amongst infected people – blueberries. In fact 17 of the 29 cases (59%) were shown to be associated with consumption of blueberries. Six blueberry samples were examined in the laboratory and were found to be contaminated with HAV. It turned out that a blueberry grower had a child who had HepA, in addition the grower had no toilet facilities near to the berry fields and so the pickers relieved themselves in the field. Human faeces were found in the field too, the rest is a bit unclear, but either the child infected the blueberries, or infected pickers, or both. The outcome was serious and required urgent attention. We had a situation in which a HepA-infected person might poo in the field, could not wash their hands, then might pick blueberries, and transfer virus from the faeces to the blueberries via their contaminated hands. In addition, a nice steaming reservoir of the virus was left at the edge of the field! (Fig. 4-2).

Fig. 4-2. Picking blueberries (*photo by Margaret Tanner*)

When blueberries are eaten raw a lot of people don't wash them because they like to see the lovely grey bloom on the fruit. In this scenario this would have increased the risk of infection. We don't know whether washing would remove all of the virus particles, but it is likely to remove some and so decrease the risk of infection from eating the contaminated fruit.

HepA in Raspberries

Fruits are a good vector for HepA and there have been several cases of their involvement in transmission of the disease. Another example was in Scotland ten or so years ago when contaminated raspberries were fed to a medical conference as a seasonal treat from a part of the world renowned for its raspberries. A significant number of the conference delegates contracted HepA ... rough justice!

Other Viruses

There are other viruses that might be transmitted to people via food but we do not regard them as conventional food borne viruses (partly because their transmission is rare, and partly because food is not their primary route of transfer). Shellfish are good vectors for many viruses because they filter their food from sea water, therefore if there is a virus (or a bacterium, or other pathogen) in the water the shellfish might filter it out along with its food. The unsuspecting consumer of the raw shellfish might contact the viral disease.

The polio virus is a good example of a water-borne virus that might contaminate sea water. But where does it come from? The most likely source is faeces dumped from boats and ships. Fortunately polio is relatively rare in the developed world and therefore it is not very likely that the virus will be present in the vicinity of shellfish fisheries. Despite this we do have worries about the possibility, especially as the polio virus can withstand some of the preservation methods for shellfish (e.g. marinating mussel). Cooking will kill most viruses, so these concerns apply only to uncooked shellfish.

A possible scenario of infection illustrates this potential problem well. In the north of New Zealand's North Island there is an idyllic place called Bay of Islands. This boatie's wonderland attracts many tourists who spend their time sailing around the bay, eating nice food, and drinking wonderful New Zealand wines. Heaven! The Bay of Islands is also an excellent shellfishery. Both feral molluscs and shellfisheries are important sources of green lipped mussels and oysters. The problem is that some of the boaties discard their untreated waste overboard – viruses and all. This is a problem from the point of view of conventional food-borne viruses such as Noro virus and HAV, but it could also be a route for other viruses to enter the food chain if one of the boaties harboured, for example, polio virus. The New Zealand authorities were sufficiently worried about the problem to close the Bay of Island fisheries and make it illegal to take shellfish for human consumption (Fig. 4-3) (see also page 86).

There are many other viruses that could be transmitted in this way, most of which we haven't even thought about.

NEW ZEALAND

Bay of Islands

Auckland

Wellington

Christchurch

Dunedin

Fig. 4-3. New Zealand showing Bay of Islands

Viruses Are Not All Bad

As I mentioned at the beginning of this Chapter, there is a very interesting family of viruses called bacteriophages or phages for short. They are the goodies of the viral world because they infect and kill bacteria. Phages are specific to a bacterial species, e.g. T4 Phage infects only *E. coli* (Fig. 4-4 and Fig. 4-5).

Phages are spectacular viruses, they are built like microscopic Lunar Landing Modules. They dock onto the surface of

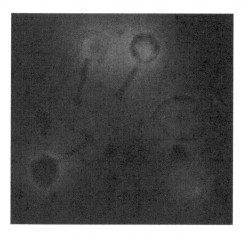

Fig. 4-4. Electron micrograph of bacteriophages. These amazing "creatures" infect cells by injecting their nucleic acid into the host cell. The phage on the left has a white head because it has injected its nucleic acid into its host electron micrograph taken by Manfred Ingerfeld at the University of Canterbury, New Zealand and kindly provided by Gwyneth Carey-Smith

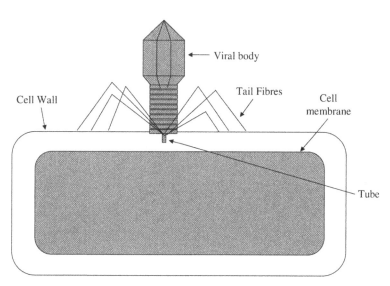

Fig. 4-5. A bacteriophage docked on a bacterial cell surface

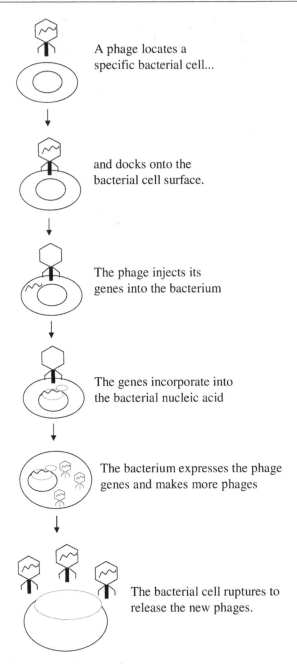

A phage locates a
specific bacterial cell...

and docks onto the
bacterial cell surface.

The phage injects its
genes into the bacterium

The genes incorporate into
the bacterial nucleic acid

The bacterium expresses the phage
genes and makes more phages

The bacterial cell ruptures to
release the new phages.

Fig. 4-6. The process of phage infection

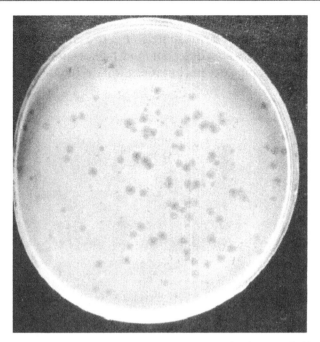

Fig. 4-7. Phage plaques on an agar plate (*photograph kindly provided by Gwyneth Care-Smith, ESR, New Zealand*)

the bacterium-using specific bacterial cell surface proteins to identify their host. Then inject their genetic material into the bacterium, the bug then replicates the virus and breaks open (lyses) to release millions of new phages that infect the surrounding bacterial cells (Fig. 4-6).

It is easy to demonstrate this effect in the lab by infecting a bacterial culture with phage. The phages infect the bacteria growing on the Agar plate and kill them creating zones of dead bacteria called plaques. You can see these in the picture (Fig. 4-7).

Scientists are just beginning to see the value of these viruses as a means of making our food safe. One of the most problematic food-borne pathogens is *Campylobacter*. If we could find a Campy phage we could infect food with the phage and let the phage kill any Campy on the food and so make the food safe. This is a brilliant idea. In a few years I think that exactly this will be done. Looking even further ahead, it might be possible to infect farm animals, or entire farms, with pathogen-specific phages so wiping out food-pathogens on the farm. This might be a pipe dream, but it is a wonderful idea … watch this space.

5
Mad Cow Disease and the Elusive Prion

5   Mad Cow Disease and the Elusive Prion

In 1986 a new disease of cattle struck the UK. At first it looked like a fairly rare, rather esoteric disease of more interest
to white-coated scientists than to anyone else. But within
5-years the disease had resulted in collapse of the UK's prestigious beef industry, a political furore that resulted in the
Brits no longer trusting what they were told about food safety,
and the rest of the world looked on with horror as people contracted the disease from their food and later died in an horrific
way with no hope of cure. This, of course, was Bovine Spongiform Encephalopathy (BSE), or Mad Cow Disease as the UK
press christened it. It changed the face of food safety and diminished public trust in politicians and scientists throughout
the world.

A New Disease Emerges

During April 1985 a veterinary surgeon, Colin Whitaker, examined a cow on a farm in Ashford, Kent in the UK. The normally
placid animal was behaving oddly and becoming hyperactive
and aggressive towards the farmer and found it difficult to control its limbs (ataxia) so it staggered about when it tried to walk.
The cow's symptoms closely resembled Staggers, a condition
known for many centuries and caused by a lack of magnesium
in the diet. Vets normally treat Staggers with magnesium, but in
this case the treatment had no effect. In fact the animal got

worse. Mr Whitaker referred the case to the Central Veterinary Laboratory (CVL) in Surrey, UK for the experts to have a look.

At first the CVL scientists and vets thought that the cow might have been poisoned, perhaps with an organophosphorus (OP) pesticide that attacks the nervous system which would explain both the behavioural changes and the animals difficulty in controlling its limbs. A great deal of work ruled out poisoning, and left the scientists with no explanation. They were almost at the point of accepting that they had been beaten by the case, but that it was probably a one-off and would have to be accepted as one of life's unexplainables. Then another case arrived, and another, and another, until, by the end of 1986, nine herds had been affected often with more than one affected animal per herd.

The incidence of the disease increased dramatically until its peak (7,267 herds affected) in 1992, with a steady decline to the latest figure of 18 herds affected at the end of July 2003 (Fig. 5-1).

By 2003 there had been a total of 180,166 cases. Nobody in their worst nightmare would have predicted an epidemic of such devastating proportions to have followed that first ataxic cow in April 1986.

What is BSE?

When the pathologists examined the ataxic cow they found that the brain looked very characteristic under the microscope. It had numerous "holes" that gave it a spongy appearance. This is indicative of a group of diseases called the transmissible spongiform encephalopathies (TSEs):

- *Transmissible*   can be transferred from one animal to another
- *Spongiform*   because of the spongy microscopical appearance of the brain
- *Ecephalopathy*   disease of the brain (*cephalos* – Greek for head; *pathos* – Greek for suffering)

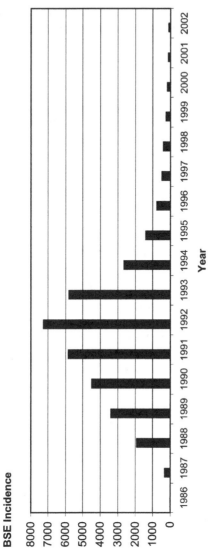

Fig. 5-1. BSE incidence in the UK (*data from the Department for Environment, Food and Rural Affairs [DEFRA], UK at http://www.defra.gov.uk/animalh/bse/bse-statistics/bse/herds.html*)

This was the first time that a TSE had been seen in a cow; it was a brand new disease about which nobody knew anything.

Other Animal TSEs

The TSEs comprise a group of quite unbelievable (their causative agent is truly fantastic, but more of this later...) diseases of many animal species, including humans.

Scrapie

Scrapie of sheep and goats was the first TSE to be discovered and has since become the most studied and reviewed of all the STEs. Until the early part of the 20th century this disease was rarely considered by vets, not because it was a new disease, in fact it was first described in 1732, but because shepherds often tried to conceal it. Scrapie is thought to have started in Spain and is now widespread in Europe, Asia and America but has never been seen in Australia or New Zealand. Trade in live animals spread the disease around the world. New Zealand's and Australia's strict biosecurity laws protected these countries.

Scrapie sheep have been found in 35 counties in England and Wales and it has been suggested that about a third of flocks in the UK are affected.

Scrapie can be transmitted within flocks by infection of lambs at birth or by other members of the flock contacting (perhaps eating) afterbirth tissue (e.g. the placenta).

Transmissible Mink Encephalopathy (TME)

TME was first identified in Wisconsin, USA in 1947 with further outbreaks in the 1960s. The symptoms and progression of the disease are strikingly similar to scrapie. In captivity mink are often fed a diet derived from animal remains, including sheep meat and bone, for this reason it was suspected that TME was due to infection of mink with the scrapie agent.

Chronic Wasting Disease (CWD) of Deer and Elk

Chronic Wasting Disease was first noted in 1967 in captive mule deer in Colorado, USA. It closely resembled scrapie and is known to be transmissible within species. In the period 1981–1995 CWD was confirmed in 49 free-range deer from North Central Colorado (USA), and later in 2002 there was an outbreak in deer in the mid-west of the USA.

Feline Spongiform Encephalopathy (FSE)

FSE was first identified in a five year old Siamese cat at the Bristol Veterinary School, UK. Retrospective examination of tissue sections from cats dating back to 1975 revealed no similar cases and therefore FSE was designated a new disease. By October 1996 75 cases had been reported in the UK. In addition, by 1997, FSE had been confirmed in 2 pumas, 6 cheetahs, 2 tigers and 1 mountain cat in UK zoos, suggesting that this is a disease of the cat family rather than being confined to the domestic cat.

Human TSEs

Creutzfeldt-Jacob Disease (CJD)

In 1920, Hans Creutzfeldt reported a new and unusual neurological disease. The following year Alfons Jacob reported four cases of progressive fatal dementia which he grouped together as examples of "spastic pseudosclerosis" and believed that they resembled the case originally described by Creutzfeldt. The disease now bears both names in recognition of their contribution. The clinical course and features of CJD have been well documented. CJD is a rare disease which is found world-wide at a rate of approximately 1–2 cases per million population per year.

Kuru

Kuru is a condition confined to highland New Guineans in the mountainous interior of Papua New Guinea and is believed to result from ritualistic cannibalism. The brain of deceased relatives was eaten by women and children and the muscle by men. The level of infectivity was greatest in the nervous tissue and this is reflected in the fact that women and children were those mainly affected by Kuru. This cannibalistic practice has disappeared over the last 30 years and by 1985 the disease was no longer seen in anyone under 35 years of age. Our knowledge of Kuru was instrumental in sorting out how BSE spreads.

Gerstmann-Strausser-Scheinker Syndrome (GSSS)

In 1928 Josef Gerstmann reported "An interesting case of hereditary familial disease of the central nervous system". After his patient's death in 1932, Gerstmann joined with Ernst Strausser and Isaac Scheinker to publish a detailed case report. GSSS is a rare disease which occurs at an order of magnitude lower rate than CJD. It only occurs in families and is closely related to CJD, it is genetically controlled and occurs in families by inheritance of a group of genes.

The scientists investigating BSE soon realised that it was a TSE and were able to call upon the vast amount of knowledge of the other TSEs to unravel this incredible new disease. In the five or so years following the discovery of BSE an unprecedented research effort resulted in a good understanding of the disease, its causative agent, and its means of transmission. It is a pity that this truly wonderful research was lost amidst the political argument that raged about the disease and its possible effects on consumers ... but more if this later.

What Causes BSE?

Stanley Prusiner had been working on Scrapie at the University of California, USA for many years. He, and others, had tried to identify the cause of scrapie.

- Was it a bacterium? No, bacteria can't survive 100°C (since then bacteria living around volcanic vents deep in the ocean that thrive at temperatures above 100°C have been discovered) – boiling scrapie extracts did not significantly reduce their infectivity.
- Was it a virus? This was a bit more difficult to decide, but most scientists thought not because scrapie extracts treated with UV or γ-radiation were still infectious and viruses are killed by these high energy rays.

So what causes scrapie? Professor Prusiner discovered a truly amazing causative agent. He called it a proteinaceous infectious particle or *Prion*. It is a protein. It is not alive. But it behaves just like any other infectious agent in that it replicates itself in the infected animal's body. Many scientists simply did not believe this apparently far-fetched hypothesis.

However the prion has stood the test of time and incredible scientific scrutiny, and now it is accepted as the causative agent of the TSEs. Prusiner was awarded the Nobel Prize for Medicine in 1997 for his remarkable discovery – his tenacity paid off.

What is the BSE Prion and How Does it Cause BSE?

Prions are medium sized proteins (molecular weight = 33–35,000 daltons), they are found in most, if not all, cells and are thought to play a role in communication and recognition between cells. They are called cellular prions (or PrP^C in scientific jargon – PrP stands for *prion protein*, and C stands for *cellular*). There are damaged forms of prions which resemble very closely PrP^C, but are different enough not to function properly – indeed they are highly dangerous. These are the so-called scrapie prions

(PrP^{SC}), but in fact are the TSE prions – they are called scrapie prions because they were identified from scrapie sheep.

So, what is the difference between PrP^C and PrP^{SC}? The simple answer is, very little. They have the same amino acid building blocks in their protein structures, they have the same molecular weight, but they have one very important difference. The shape of their protein make-up is different. Proteins are made up of long strings of amino acids that are folded to make complex structures. The folding (i.e. conformation) of PrP^C and PrP^{SC} is different. This apparently small difference in shape makes an enormous difference to their biological activity – this story is almost incredible, I never cease to marvel at the ingenuity of the prion as a disease causing agent.

Protein biochemists have given names to the different types of protein molecule folding. PrP^C has a lot of α-helix, PrP^{SC} has more β-pleated sheet. α-Helix proteins look like spiral staircases (well they would if it were possible to magnify them enough), β-pleated sheet proteins look like stacks of plates or folded sheets (Fig. 5-2).

Now comes the amazing bit. If a molecule of PrP^C comes into contact with a molecule of PrP^{SC} the PrP^C is pulled into the same molecular shape as the PrP^{SC}. This is called an induced conformational change because one molecule has induced a change in shape (conformation) of another. This, of course, has significant implications because it means that dangerous PrP^{SC} can be created from safe PrP^C. The implications become even more worrying when the PrP^{SC} is in the brain next to PrP^C molecules doing their important job helping cells to communicate with each other. The multiplication of PrP^{SC} in this way is devastating to the brain's function. It stops cell to cell information flow and causes very serious brain malfunction (Fig. 5-3, see also p. 111).

Are Prions Alive?

When scientists first saw the PrP^{SC} replication process they thought that the prion was dividing and growing like a virus. This is not the case. The prion is simply a chemical that is able

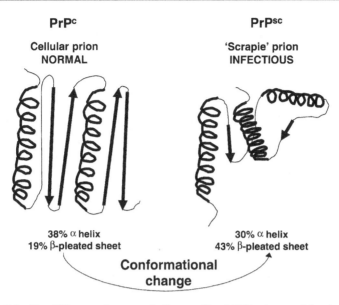

PrP^c

Cellular prion
NORMAL

PrP^{sc}

'Scrapie' prion
INFECTIOUS

38% α helix
19% β-pleated sheet

30% α helix
43% β-pleated sheet

**Conformational
change**

Fig. 5-2. The differences between the "normal" and BSE prions and the shape change that leads to BSE – the curly line represents α-helix protein; the arrows represent β-pleated sheet protein

to reproduce its form by changing other molecules into the same shape as itself. It is a very devious poison.

What Happens if you Eat Beef Infected with BSE Prion (PrP^{SC})?

In the early days of our understanding of TSEs it was thought that nothing would happen. After all Scrapie was first described in 1730s, many people have eaten infected lamb since then, but nobody had contracted Scrapie. But humans get CJD, could this be the human form of Scrapie? Studies showed that Scrapie and CJD are not connected – CJD is caused by a spontaneous change (mutation) in the gene that codes for human PrP^C. The mutation causes the synthesis of a CJD prion rather like PrP^{SC} that has just the same devastating effect.

The UK government took this information to heart and were unable to move with changing scientific ideas. Partly be- _____

cause they did not want to believe the evidence that was beginning to appear, and partly because the scientists advising them were uncertain – this was cutting edge science in a field that was almost unbelievable, and the implications would be devastating to the UK's farm economy.

It was becoming clear that the BSE prion was different to the Scrapie prion. It behaved differently. Might is jump species? Might human consumers contract it if they ate meat from a BSE cow? These were real scientific questions, based on uncertainty, but they had no answers. A great deal of very expensive research was necessary to answer them. But there was no time – answers were needed NOW!

The press got hold of the possibility that people might catch BSE – a furore of unprecedented proportions broke out. The Brit's were scared. Would they catch this terrible disease? Would everyone who had eaten beef since the onset of BSE die a terrible death? There was confusion and fear.

Then came the worst possible news. Researchers had shown that a new form of CJD, which they called new variant CJD (*nv*CJD) was caused by eating BSE-infected meat. In the UK at this time, there was that terrible feeling of a lull before an enormous storm. Scientists and medics alike were shocked, silent, even frightened. What were the implications? Would our worst fears be realised?

nvCJD

*nv*CJD or *v*CJD as it is now usually called, first appeared in England in 1995/96. The cases were very like CJD, but unlike CJD they occurred in young people – CJD is a disease of later life, not usually occurring before 50 years old. The peculiar CJD cases were in people in their teens and 20s. The first case was considered a curiosity. Then came another, and another – just like BSE in cattle. By 1996 there were 5 early onset CJD cases. They had strikingly similar symptoms.

The first case was in a dairy farmer, then came a teenager, then a 28 year old woman. Their symptoms were loss of memory, confusion, mood changes, difficulty walking, loss of coordination, dementia, and death. A terrible scenario, horribly reminiscent of BSE.

All of the early onset CJD (as it was then called) patients died within 14 months of the onset of symptoms (this is different to CJD where the period to death is only 4 months) and were found to have identical brain microscopical appearance at post mortem examination. They looked strikingly similar to BSE.

Since the first cases in 1995, there has been a meteoric rise in cases, until a peak occurred in 2000. Then followed the long awaited decline (Fig. 5-4).

Cases of vCJD have now been recorded in other countries. Most have been traced back to consumption of UK beef.

How Does the Prion Get to the Brain?

If meat contaminated with the BSE prion is eaten the prions pass through the stomach unscathed (they can withstand stomach enzymes that break down proteins), are absorbed across the intestine – like other food components, and find their way to the spinal cord (possibly via the lymphatic system). When they reach the spinal cord they slowly move up towards the brain. They are not alive remember, so they don't propel themselves in any way. They simply diffuse like any other chemical. This process is slow. It takes many years (this is the "incubation period" of the disease); in humans this probably takes about 10 years (the first case of vCJD was seen in 1995, and the first case of BSE was in 1986, so the first human exposures were also in 1986; i.e. 9 years from exposure to onset).

When the prion gets to the brain it meets brain PrPC and converts it to PrPSC and there is no going back. vCJD and death are inevitable. It is likely that the concentration of BSE prion in the consumed meat is a very important determinant of vCJD risk. Nobody knows how much BSE prion is needed to result in vCJD, but it is thought that only large doses are certain to cause the disease. This probably explains why the incidence of vCJD was so low compared to the theoretical exposure of consumers

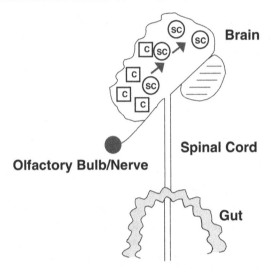

Fig. 5-3. The BSE prion replicates in the brain by changing the shape of normal prions that it comes into contact with

to BSE beef between 1986 and 1989 when the Offals Ban was introduced.

What Did the UK Government Do to Minimise BSE Risk to the Consumer?

Let's back track a little. The first case of BSE was confirmed in 1986, but it was not until 1996 that the first case of vCJD was confirmed and the link between the two diseases confirmed. So, before 1996 there was no evidence that BSE could affect human consumers. I point this out because it is very easy with hindsight to criticise the action of the UK government in the way that they minimised the risk that they knew nothing about.

As research on BSE unravelled, what causes the disease and how it was transmitted, it became possible to introduce risk management strategies to minimise human exposure. The first knee jerk reaction was to ban eating beef. This was a ridiculous suggestion in a nation where roast beef and Yorkshire pudding is the national dish (the French even affectionately call – I think!

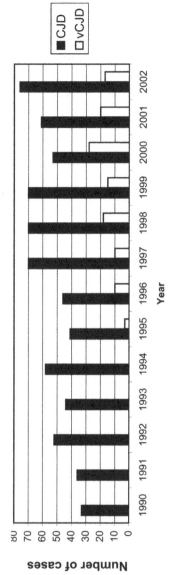

Fig. 5-4. The incidence of CJD and vCJD in the UK showing the long awaited decline that began in 2001 (*data from the CJD Surveillance Unit, Edinburgh, UK. See http://www/cjd.ed.ac.uk/figures.htm for up to date figures*)

– the Brits "rostbifs"). But it was a starting point for the risk managers to grapple with. Research had shown that the BSE prion is at highest concentration in the CNS – so the first risk management approach was to ban eating brain and spinal cord in 1989 (the Offals Ban). You might think that not many people eat the CNS, but in fact most of us did. Brain and spinal cord were included in the mixtures of meat used in making sausages, meat pies, beef burgers, etc. Since the Offals Ban was not introduced until 1986, the Brits (and visitors to the UK) had been exposed to BSE meat for 3 years.

As we learnt more about the BSE prion and where it occurs in the cow's body, other tissues were included in the Offals Ban (e.g. parts of the intestine). Milk was shown to be safe and other tissues (e.g. muscle) were very low risk.

The Offals Ban was a good risk management strategy that worked well. We did not know this at the time, but looking at the vCJD statistics show us that exposure between 1986 and 1989 probably resulted in the 136 human deaths that have occurred so far (up to September 2003 – there will be more, but not many). The Offals Ban very significantly reduced human exposure to the BSE prion which explains the onset of vCJD decline in 2001. If risk management had not been introduced we would have had a vCJD epidemic of frightening proportions. The sadness, of course, is that anyone had to die, but due to the incredible science that was conducted over a ridiculously short period of time, I am able to report than hundreds rather than 10s of thousands of people died. We should congratulate the scientists, not ridicule them.

Since 1996, there has been a steady accumulation of evidence, often tragic, that supports the vCJD from BSE beef theory. Perhaps the most heart-wrenching was a cluster of vCJD cases that appeared in the Queniborough, Leicestershire, UK in between 1998 and 2000. Five people died of vCJD which is an incidence very much higher than would be expected for such a small village (population 1,800) on the basis of the national statistics (for this population even a single case of vCJD is a very remote possibility). It turned out that a local butcher used traditional butchery methods that might have led to cross contamination of meat from a BSE cow. The butcher's knives were suspected; perhaps an infected brain was cut, then the same knife used to cut some meat.

Does Cooking Destroy the BSE Prion?

The BSE prion is an incredibly resilient protein. As discussed above, it can withstand enzymes that destroy most proteins, it is acid and base stable, and is not destroyed by normal cooking temperatures. At 138°C it begins to lose its activity. This temperature might not sound very high, but most meat only reaches 60–70°C in its centre during cooking. If you made sure that your roast beef was kept at 138°C for long enough to destroy the prion, you would end up with a solid charred mass. Safe (perhaps) but inedible! Therefore how ever hot you cook your food the BSE prion will survive.

It is interesting that if you look at CJD incidence statistics (from before we knew about vCJD) you will see that medics and paramedics have a greater incidence than the general population. This is because neurosurgeons sometimes cut themselves accidentally with their surgical instruments. If they have operated on a CJD-infected patient, who has not yet begun to show the symptoms of the disease, they might infect themselves with the CJD prion. The normal sterilising procedures used in hospitals use pressurised steam at 121°C (autoclaving) is not hot enough to deactivate the prion. When this was realised, new procedures were introduced to sterilize surgical instruments used in CNS surgery – they now have to be dry heated in an oven at 140°C to be sure to destroy infectious prions.

Why Was the First Case of BSE in 1986?

This is a key question. What happened, or changed, in farming practice to cause the disease. Or was it just a fluke?

Prions are passed from animal to animal orally (including humans, remember Kuru). If a non prion-infected animal

113

The incidence of CJD in selected* occupations in the UK between 1990 and 1995. This shows that medics and paramedics are more likely to die of CJD than non-prion exposed groups. The data are from the annual report of the CJD Surveillance Unit, Edinburgh, UK (1995)

Occupation	CJD Incidence/million
Managers	0.3
Secretaries/clerical workers	0.5
Shop workers	1.7
Medics/paramedics	5.7
Teachers	0.7

* Some data have been omitted because there is not sufficient space in this section to discuss in detail the higher incidences in some other professions.

eats a prion-infected animal it is likely to contract the prion disease (e.g. BSE). But cows don't eat cows ... or do they? It is true that cows are not carnivorous by nature, but farming developments have forced them to eat each other by using meat and bone meal (MBM) as a component of cattle feed. This began a long time ago, and was a way of putting back valuable elements (e.g. calcium) into farm animals' food chains.

When farm animals are slaughtered, their meat is removed from the carcass, bones, plus some attached meat are left. This is treated to remove the fat – the product is tallow used to fry food in. The remaining material is dried and ground up to make MBM. This used to be added to animal feed to prevent waste and to get rid of a waste product that otherwise would be of very little use. This is the ultimate in recycling.

Until the early 1980s in the UK, the tallow was extracted from the animal remains after the meat had been recovered using solvents. The solvents were then evaporated off to leave the fat which was sold to fish and chip shops and the like. The remainder was dried and ground to make MBM. In the 1980s the process for recovering tallow was changed. Instead of solvents, heat was used. The carcass minus its meat was passed through an oven and the tallow melted off. This removed the need for solvents and allowed the process to be streamlined.

It is thought that the solvents inactivated the BSE prion, but the heating process did not. So, the change in process allowed the BSE prion to survive and infect cattle via the MBM in their feed.

This still does not explain fully where BSE came from. Some very elegant work at the Central Veterinary Laboratory in Weighbridge, UK came up with the answer.

The Origins of BSE

The first few cases of BSE were all associated with a single animal feed producer in the south of England. It was thought that scrapie sheep MBM had been used to make cattle feed and that the scrapie prion had jumped the species barrier and infected cows. The cows then formed a new batch of MBM, which was fed to more cows so amplifying the disease. This mechanism of infection would very quickly result in a widespread epidemic. The Scrapie hypothesis was accepted for quite some time, but it did not quite ring true. Scrapie is a common disease and so why did only one feed producer transmit the prion? This simply did not make sense. A better, and now widely accepted, explanation is that a prion mutant cow – i.e. a cow making defective prion (PrP^{SC}) was produced randomly due to a spontaneous and 'natural' change in the gene (mutation) that codes for the prion protein, this cow was made into MBM and incorporated into cattle feed. This is a far better explanation of the point source of infection. So the fact that BSE occurred in the UK was just damn bad luck. The mutation could have occurred anywhere. Indeed such mutations might happen from time to time, which might explain the few BSE cases around the world that cannot be traced back to UK feed or UK breeding stock. So countries that say they are BSE free should beware – nature might well prove them wrong!

Can BSE Be Transmitted from Cow to Cow?

There are only two ways that the BSE prion can be transmitted from animal to animal:

- Horizontally – by a cow eating BSE-infected MBM
- Vertically – at birth a calf can be infected from its BSE infected mother

It is not possible for one animal, or a person for that matter, to catch a TSE from another animal or person in any other way. In people, vCJD is transmitted from cow to human horizontally, whereas Kuru is transmitted from person to person horizontally. There is no known vertical transmission.

The Politics of BSE

The risk of getting vCJD from BSE-infected beef is very low indeed. There have been 136 vCJD deaths in the UK out of a population of about 60 million. Not all of the population eat beef, and so some could never have been exposed to the BSE prion. If we assume that 40 million Brits eat beef (in reality it is likely to be more), if they were all exposed in the same year (which of course they were not), the risk of "catching" BSE is 1 in 294,000. This is small, but the real risk is much smaller because exposure was over at least 4 years (i.e. the period before the Offals Ban). Despite the relatively low risk, consumers around the world shunned UK beef and beef products. The UK beef industry collapsed.

Politicians were desperate to support this important industry and so hawked the "there's no risk to the consumer" story. This was stupid. They were doomed to be proved wrong. There is no such thing as no risk. It would have been better to explain the issues to the public, but they were immensely complex, and the science was unravelling as the news stories broke. The result was a series of knee jerk reactions, contradictory stories, changed advice – all of which undermined the public's confidence in both politicians and scientists. It will take many years to heal this wound.

A look at some of the newspaper headlines published around the world throughout the Mad Cow Disease saga says it all.

Beef brains banned in food
The Times, 14 June 1989

Mad cow disease 'now in decline'
The Independent, 23 September 1993

Human 'Mad Cow' deaths at new high
The Daily Telegraph, 6 October 1995

Death toll from brain disease increases
The Times, 8 October 1995

Top scientist adds to BSE warnings
The Independent, 4 December 1995

Why we should all give up beef
The Independent, 7 December 1995

Food giant may sue BBC in beef scare
The Independent, 7 December 1995

Lax regulations to blame for BSE, says new study
The Independent 12 December 1995

Expert warns of epidemic
Evening Post [Lancashire, UK]. 21 March 1996

Latest BSE scare puts farming in crisis
Farmers Guardian, 22 March 1996

Beef industry faces ruin as bans spread
The Times, 23 March 1996

Mafia linked to sale of herd infected with BSE
The Times, 27 March 1996

Birds Eye stop making burgers
The Times. 27 March 1996

Food firms try to cut all sources of British beef
The Times, 27 March 1996

Businessman, 42, latest suspected victim of CJD
The Times, 3 April 1996

Cattle slaughter may be doubled to 30,000 a week
The Times, April 12 1996

Laboratory mice carry secret of human BSE risk
The Times, 25 October 1996

We won't swallow any more lies about food
The Independent, 31 January 1997

CJD kills five around village
The Sun, February 10 2000

Transfusions can spread BSE
The Times, 15 September 2000

Madness in the blood
The Sunday Times, 24 September 2000

Mad-cow fears prompt EU meat ban
The Press [Christchurch, New Zealand] 8 January 2001

Warning of second wave of vCJD
New Zealand Herald, 16 May 2001

BSE sold a great many newspapers, destroyed thousands of farmers' livelihoods, lives seriously impacted on the global beef industry, has killed over a hundred innocent consumers, ended the careers of a handful of politicians, but led to some of the most wonderful science of our time. This was truly a disaster, but it was a triumph of discovery too.

STANLEY B. PRUSINER
1997 Nobel Laureate in Medicine – *For his discovery of Prions – a new biological principle in infection.*

6
Natural Toxins in Food

6   Natural Toxins in Food

There is much concern about horrible man-made chemicals that find their way into our food. Pesticide residues in fruit and vegetables, veterinary medicines in meat, nitrate in lettuces, etc, etc (see Chapter 7). Most people think that man-made (artificial) is bad and natural is good. This is rubbish! Some of the most toxic chemicals that we know are natural, so perhaps we should think again.

Just to set the scene, here are some natural and artificial chemicals with their toxicities – (i.e. rat LD50s – remember this is a measure of toxicity the smaller the number the more toxic).

The table of toxicities (LD50s) of natural compared with man-made poisons shows clearly that natural is not necessarily safe! In fact, it is abundantly clear from this very short list that some natural toxins pose a far greater hazard than artificial (man-made) chemicals.

Chemical (source)	Rat oral LD50 (mg/kg body weight)
For reference, a commonly consumed, accepted poisons:	
Ethanol (the alcohol present in beer, wines, and spirits)	7,000
Aspirin	1,240
Artificial chemicals that might contaminate food:	
Diazinon (an organophosphorus insecticide)	250
Glyphosate (*Round-up* – a herbicide)	4,873
Atrazine (another commonly used herbicide)	1,750*
Penicillin-G (an antibiotic)	6,900
Natural chemicals in food:	
Tetrodotoxin (from Fugu Fish, a delicacy in Japan)	0.01**
Solanine (from potatoes)	42**
Psoralen (from parsnips and related plants)	791

* Oral LD_{50} in the mouse.
** Intraperitoneal (i.p.) LD_{50} in the mouse – i.p. is injection into the abdominal cavity, it resembles oral dosing metabolically.

The value with a single asterisk is the oral LD50 in the mouse and the values with two asterisks are intraperitoneal (i.p.) LD50s in the mouse – i.p is injection into the abdominal cavity, it resembles oral dosing metabolically.

Why Do Plants Have Natural Toxins?

Plants might be eaten by animals or infected by bacteria and fungi. The presence of natural toxins helps the plant to protect itself from such attacks. For example North American Milkweed (*Aesclepias eriocarpa*) contains a very interesting, highly potent collection of natural toxins:

- Eriocarpin – LD50 (mouse)   =6.5 mg/kg body weight
- Labriformidin – LD50 (mouse) =3.1 mg/kg body weight
- Labriformin – LD50 (mouse)   =9.2 mg/kg body weight

Just 44 mg of eriocarpin will kill a sheep. This protects the Milkweed from grazing animals. However, the Monarch Butterfly (*Danaus plexippus*) is not affected by the poisons and lays its eggs on Milkweed – its caterpillars eat the plant and accumulate the natural toxins in their bodies so becoming toxic to any animal that might decide to eat them. This is a clever use of natural plant toxins in insect protection, and illustrates the importance of these highly toxic chemicals.

Other natural plant toxins prevent insect attack – i.e. natural insecticides, and others are natural fungicides. Many of the toxins have unknown functions, perhaps they are accidents of plant evolution – perhaps they just scare off grazing animals because they taste bad.

In general, leaves, roots, and tubers are much more likely to contain natural toxins than fruits, but this is by no means always the case, (e.g. potato fruits contain very toxic solanines). The seeds in fruits are the means by which plants reproduce, and often they rely on animals eating the fruit as part of the seed transmission mechanism. The seeds might pass through the animal and be deposited with a dollop of fertilizer onto the ground, or the animal might discard the seeds which drop onto the ground beneath where they are eating. It is therefore not in the plant's interest to poison the vector of its seeds. On the other hand the leaves, roots and tubers are important to the plant in a different way. Leaves photosynthesise (i.e. make sugars from carbon dioxide using sunlight energy), roots take up water and nutrients, and tubers store nutrients for the dormant months and to allow rapid growth in the spring. Clearly the plant does not want animals to eat these important organs, or microbes to infect them. It is for this reason that the plant might produce natural toxins, to ward off this attack.

There are many natural toxins that we are exposed to everyday in the plants that we eat. The dose that we receive is very low and therefore even though the toxins have a very high hazard (e.g. low LD50) our exposure is low and so the risk is also low. However, from time to time the levels of natural toxins in plants might increase and so the consumer could get a larger dose and become ill.

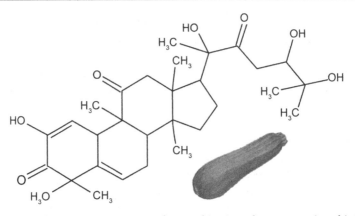

Fig. 6-1. The molecular structure of a cucurbitacin and a courgette (zucchini) the source of this deadly poison

Cucurbitacins in Courgettes (Zucchini)

In the early summer of 2001, several New Zealanders complained of stomach cramps after they had eaten zucchini. Over a few weeks more and more cases were reported, which led the health authorities to investigate. Many of the people who became ill remembered that the zucchini tasted bitter – I'll return to this important fact later.

Members of the cucumber family (Cucurbitacea) are able to produce a group of highly potent toxins (cucurbitacins) that have insecticidal and/or fungicidal properties. The production of cucurbitacins is controlled by the plant so that they are only made when they are needed. In fact the gene that codes for cucurbitacin is only switched on if the climatic conditions are right for insect infestation or fungal infection. The weather in New Zealand in the early summer of 2001 was just right for insects and fungi – wet and cool, the gene switched on the synthesis of cucurbitacin.

The cucurbitacins are intensely toxic (cucurbitacin-B oral LD50 [mouse]=5 mg/kg body weight – 300 mg could kill a human) and taste very bitter indeed. The cucurbitacins have such a terrible taste that it is very unlikely that anyone could stand to eat a zucchini containing enough cucurbitacins to cause them significant harm (Fig. 6-1).

Rhubarb and Oxalic Acid

Most people know that rhubarb leaves are poisonous, and that rhubarb itself is a good laxative. But why? One chemical is responsible for both properties, oxalic acid (oral LD50 [rat]= 375 mg/kg body weight – 25 g could kill a human). Rhubarb contains about 1% oxalic acid. It takes a lot of oxalic acid to kill a human, but its sub-lethal effects are seen at very much lower doses. It binds calcium to form calcium oxalate which causes an ionic imbalance in the cells of the gut and results in diarrhoea. When you eat rhubarb your teeth sometimes take on a rough feel if you run your tongue over them, this is because the oxalic acid is attracted to the calcium of your teeth.

If cream, ice cream, or custard made with milk is eaten with rhubarb, insoluble calcium oxalate is formed with the calcium in the milk products, this stops it being absorbed so reducing the effects of the rhubarb.

Glycoalkaloids in Potatoes

The glycoalkaloids are highly toxic components of members of the potato/nightshade family (Solanacea). Their concentration varies very much indeed between species. They are at highly toxic levels in members of the nightshade genus – this is one of the reasons that Deadly Nightshade (*Atropa belladonna*) is deadly. However they also occur in plants that we eat, the most notable being potatoes and tomatoes, but they are usually at non-toxic concentrations in the parts of the plant that we eat.

The three main glycoalkaloids are α-solanine (Fig. 6-2), α-chaconine, and solanidine. They are very toxic.

	LD50 [rat] mg/kg body weight
α-Solanine	42 [oral]
α-Chaconine	84 [i.p]
Solanidine	590 [oral]

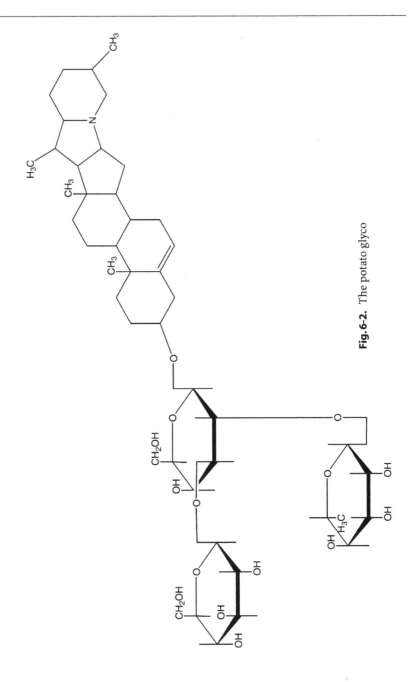

Fig. 6-2. The potato glyco

About 2.5 g of α-solanine could kill a human. This means that a meal of 1.5 kg of potato sprouts (see the table below showing levels of glycoalkaloids (e.g. solanine) in different parts of the potato plant–data from Inherent Natural Toxicants in Food (1996), MAFF, London) would be needed to be fatal, this is ridiculous and therefore it is extremely unlikely that anyone would die from eating solanine-containing potatoes. However the toxic effects of the glycoalkaloids occur at doses many times below the lethal dose.

Levels of glycoalkaloides (e.g. solanine) in different parts of the potato plant (data from Inherent Natural Toxicants in Food (1996), MAFF, London)

	Total glycoalkaloid concentration (mg/kg)
Tubers	12–20
Leaves	30–1,000
Sprouts ("eyes")	2,000–4,000
Potato skin	300–600

The glycoalkaloids taste very bitter indeed. Sometimes potatoes taste bitter, especially if they have been stored in the light; this is because in the light the potatoes synthesise glycoalkaloids. They also turn green because they produce chlorophyll – the green pigment found in leaves, it is responsible for capturing sun light energy and making sugars from carbon dioxide. Levels of glycoalkaloids in green (bitter) potatoes can be very high indeed.

Glycoalkaloid in green potatoes (data from Inherent Natural Toxicants in Food (1996), MAFF, London)

	Total glycoalkaloid concentration (mg/kg)
"Normal" potato	12–20
Green tuber	250–280
Green skin	1,500–2,200

But you would still need to eat about a kilogram of green potato skins to kill you. This is not possible at one sitting.

It is still possible to get a dose of glycoalkaloids that will cause a tummy upset. In fact skin-on potato chips (crisps) originating from green potatoes can have enough glycoalkaloid in two standard packets to result in toxicity in children – and it is not outside the bounds of possibility that a kid will eat two packets in quick succession! The table below shows the concentrations.

Glycoalkaloid in potato chips (crisps) (data from Inherent Natural Toxicants in Food (1996), MAFF, London)

	Total glycoalkaloid concentration (mg/kg)
Peeled potato chips	40–150
Skin-on potato crisps	40–720

Effects of Glycoalkaloids

The main effect is relatively mild gastrointestinal upset, although there have been more serious cases. In London in 1979 a large number of children from a school suffered from stomach pain, vomiting and diarrhoea after their lunch. An investigation revealed that potatoes from the school's kitchen had glycoalkaloid levels of 330 mg/kg. All of the kids recovered but several of them needed hospital treatment. Potatoes with levels of glycoalkaloids above 200 mg/kg are now regarded as unsafe to eat.

What Does Cooking Do to Glycoalkaloids?

The simple answer is very little. They are very heat stable. Not even frying temperatures destroy them.

Furocoumarins and Parsnips, Celery and Parsley

The furocoumarins are a group of chemicals that occur naturally in a wide variety of plants, but are at their highest concentrations in members of the Umbelliferae family (having flowers like an umbrella – parsnips, celery, parsley, etc). But they are also found in citrus fruits and figs.

There are many different furocoumarins, but they all have very similar molecular structures (Fig. 6.3).

Furocoumarins in commonly eaten fresh foods (data from Inherent Natural Toxicants in Food (1996), MAFF, London)

Plant/part	Main furocoumarin	Concentration (mg/kg)
Umbelliferae		
Celery/stalk	Bergaptan	1.3–47
Parsnip/root	Bergaptan	40–1,740
Parsley/leaf	Isoimperatorin	11–112

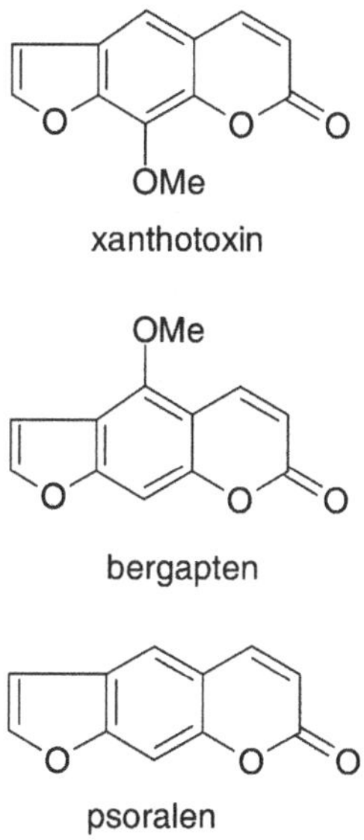

xanthotoxin

bergapten

psoralen

Fig.6-3. Molecular structures of some furocoumarins found in celery, parsnips and parsley (drawn by Barbara Thomson, ESR, New Zealand)

What Do Furocoumarins Do for the Plant?

They are produced in response to stress (e.g. bruising) and therefore are thought to prevent fungal attack. Some might also have insecticidal properties and could be produced in response to insect attack.

Interestingly, organically grown vegetables often have higher levels of furocoumarins. One explanation for this is that conventional growing techniques use insecticides, and organic methods prohibit the use of insecticides. The conventional parsnips therefore do not suffer insect attack because the insects are killed by the insecticides. On the other hand the pesticide free organic parsnips are attacked by insects and produce their own insecticides (or fungicides to prevent microbial infection of the wound caused by the insect). Therefore organic produce is likely to have higher levels of furocouarins than conventionally grown crops (Fig. 6-4).

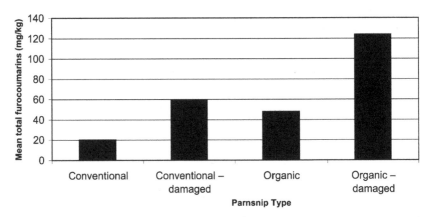

Fig. 6-4. Furocoumarin levels in parsnips showing that organic parsnips are very good at producing furocoumarins in response to damage (data from Inherent Natural Toxicants in Food (1996), MAFF, London)

Furocoumarin Toxicity

These are nasty compounds. They are activated by light (photoactivated) to form carcinogens. Therefore prolonged doses might cause cancer. They can also cause skin sensitisation to UV light – i.e. if you consume enough furocoumarins and sit out in the sun you will get a skin rash. It is not thought that normal intakes from food will cause photosensitisation, but high level intake (e.g. by people who eat large quantities of organic parsnips) might get close. This is a 'slap in the face' for organic food. I'd much rather eat a low level pesticide residue than get a dose of a carcinogen with my parsnips (Fig. 6.5).

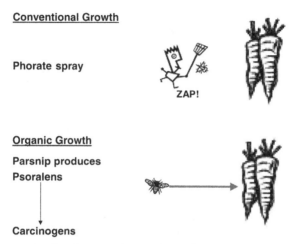

Fig. 6-5. Conventional vs organic parsnip production. Conventional growers might use man-made pesticides (e.g. Phorate) to control insects, organic farmers would shudder at the thought, so their parsnips have to defend themselves, e.g. by producing natural insecticides such as the psoralens the problem is that psoralens are carcinogenic

What Does Cooking Do to Furocoumarins?

They are not affected much by cooking. However the furocoumarins are water soluble and therefore if furocoumarin-containing vegetables are cooked in water (e.g. boiled) the levels in the vegetables will go down, but the cooking water will contain the leached furocoumarin – so if you use the vegetable water to make your gravy, you will still consume the furocoumarin.

Phenylhydrazines and Mushrooms

It is well known that many mushrooms and toadstools contain toxic chemicals. There are numerous examples of people picking mushrooms and accidentally collecting a toxic species and succumbing to its toxicity. Perhaps the best example is the Death Cap (*Amanita phalloides*) mushroom. This is amongst the most toxic plants in the world. It contains the liver toxin phalloidin which is intensely toxic (LD50 [i.m., mouse]= 0.003 mg/kg – i.m. means intra-muscular injection, i.e. injection into the muscles. 0.2 mg could kill a human), tiny doses will cause liver failure and death. There is likely to be enough poison in a single Death Cap mushroom to be fatal. The problem is that death caps look similar to field mushrooms (*Psalliota campestris*), and from time to time people make mistakes.

Phalloidin is just one of many horrific mushroom/toadstool toxins. But providing we don't eat the nasty mushrooms we'll be alright. Or will we? Shop-bought mushrooms (usually *Psalliota campestris*) also contain toxins, albeit far less acutely toxic than phalloidin, but worthy of a thought or two. One of these toxins is called agaratine (after the mushroom genus *Agaricus* from which it was first isolated). Agaratine itself is not of any great concern, however it is metabolised in the body to the 4-hydroxymethylbenzenediazonium (HMBD) ion, and this is a potent carcinogen.

There are no data on the toxicity of agaritine from mushrooms in the diet of humans, it is probably just another carcinogen in our food that plays its part in the cancer incidence rate that humans suffer – one in four of us will get cancer, there are a myriad chemicals that we are exposed to every day that contribute to this risk, agaritine is likely to be a very minor risk factor.

Herbs and Spices

Herbs and spices are used in small quantities to add flavour to our food. Some of them are quite toxic if eaten in large quantities, not that most people would want to eat large amounts of them. Some of them contain interesting chemicals that have potent pharmacological properties, indeed some of these chemicals are components of medicines or are used as herbal remedies. This is where food and medicines coalesce, there is much discussion at present about when a food is a medicine because medicines are regulated differently to foods...I'll return to this later. Meanwhile, back to herbs and spices.

Cloves

Cloves are a good example of a spice that contains a pharmacologically active chemical, indeed it was once commonly used in medicine. Cloves are the dried flower buds of a small tropical tree, *Syzygium aromaticum*. They give apple pies and cakes a wonderful flavour, and impart that very characteristic warming smell, but if you chew on a whole clove you will find a very different side to their character. They contain eugenol, not only does this impart their characteristic taste and smell, but it is also an anaesthetic. Clove oil was used by dentists as an anaesthetic until quite recently; some people perhaps still use it. So when you chew on a clove you will feel your mouth get numb. Despite its pharmacological effect eugenol is of remarkably low toxicity (oral $LD50$ [rat]=3,000 mg/kg body weight).

Eugenol is used commercially in the manufacture of vanillin – the chemical that gives vanilla its characteristic flavour and smell. It is interesting how so many of the herbs' and spices' flavour chemicals are related (Fig. 6-6).

Nutmeg

Nutmeg is the fruit of a large tropical tree (*Myristica fragrans*) – rice pudding or egg custard without a good grating of nutmeg on top to give that beautifully aromatic skin is a travesty!

Nutmeg has its fair share of pharmacologically active chemicals, amongst them, pinene, camphene, dipentene and trimyristin – they are used as perfumes, flavouring agents, and

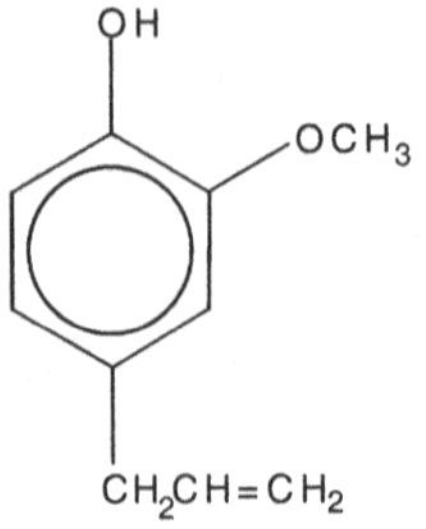

OH

OCH₃

Eugenol from cloves

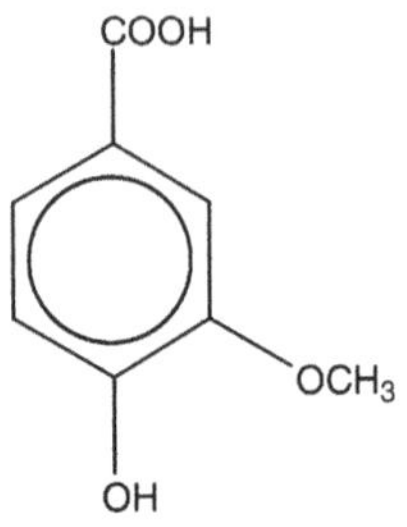

CH₂CH=CH₂

COOH

OCH₃

OH

Vanillin from vanilla

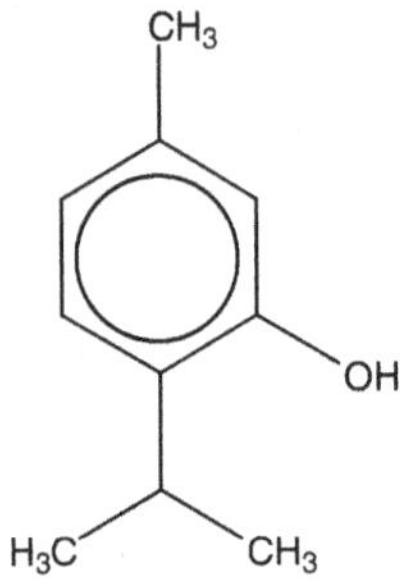

CH₃

OH

H₃C   CH₃

Elemicin from nutmeg

H₃CO

CH₂

H₃CO

OCH₃

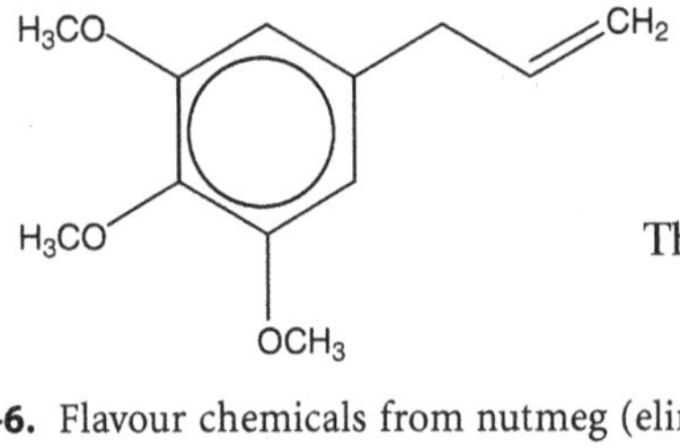

Thymol from thyme

Fig. 6-6. Flavour chemicals from nutmeg (elimicin), cloves (eugenol), thyme (thymol) and vanilla (vanillin) showing their similarities. It is interesting to speculate that the similarities might be because the molecules fit into tongue flavour receptors to send a flavour message to the brain. The receptors are like locks and the flavour chemicals act as keys to unlock the flavour signal it is just possible that the keys have to be similar in shape to fit similar receptors

trimyristin has been used to treat rheumatism. Nutmeg also contains the hallucinogen, elemicin. However, it would take a lot of elemicin to get you "high", a helping of mum's rice pudding is certainly not going to do the trick.

Thyme

Thyme (*Thymus vulgaris*) is evocative of roast lamb – it is wonderful pressed with cloves of garlic into deep cuts in a leg of lamb before roasting. It has a characteristic smell and flavour, both due in part to one chemical, thymol. Thymol is a very low toxicity (oral LD_{50} [rat]=980 mg/kg body weight) antiseptic often used in dental mouth washes, in times gone by thyme itself was used to dress wounds because of its antiseptic properties.

Sage

Sage (*Salvia officinalis*) is a herb of old English gardens. It smells lovely and has attractive blue flowers that butterflies like. In the kitchen it is used with breadcrumbs and onions to make sage and onion stuffing for poultry and pork, amongst a myriad other uses. It contains cirisiliol, a potent inhibitor of an enzyme (arachidonate, 5-lipoxygenase) involved in metabolising fats – there is some suggestion that it might protect against prostate cancer.

Chives

Chives (*Allium schoenoprasum*) are just very small onions. They contain the same flavouring chemicals found in all members of the onion family. Perhaps the most important of these is, allicin – I'll discuss this under *garlic* because the levels are much higher in garlic. There are a number of other related chemicals, all have a common chemical group – the sulphydryl (–SH) – this class of chemicals are good antioxidants, and might have other medicinal properties. Perhaps more important to the chef, they smell and taste wonderful. On the negative side, a number of them cause eye irritation and tear production (they are lachrymatory) which is why you often cry when peeling onions. A good example is diallyl sulphide which smells strongly of onions or garlic and is a powerful eye irritant, but also has anti-cancer properties. The chemical structure of allyl sulphide from onions and garlic is shown below:

$$H_2C=CH-CH_2-S-CH_2-CH=CH_2$$

Garlic

Garlic (*Allium sativum*) is another member of the onion family. As discussed above it has many of the smells and flavours of other onions, but in different concentration ratios which is why it tastes different to onions and chives. Garlic oil contains very high concentrations of allyl disulphide (note this has 2 sulphur atoms, allyl sulphide only has 1), and it is responsible for garlic's powerful flavour. Interestingly it has insecticidal properties which might be why the plant produces it. The chemical structure of allyl disulphide the chemical behind garlic's wonderful flavour is shown below:

$$H_2C=CH-CH_2-S-S-CH_2-CH=CH_2$$

Garlic is credited with many beneficial properties. It is said to be antibiotic, lower blood cholesterol, reduce blood clotting, and have anti-cancer properties. Allicin is pharmacologically active and might explain some of these possibilities, although high doses – higher than you would get from eating garlic, are usually necessary for pharmacological effectiveness in animal experiments (Fig. 6.7).

Vanilla

Vanilla is the cured unripe pod of several species of tropical climbing orchids (*Vanilla planifolia* [from Central and S. Amer-

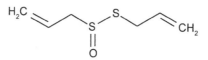

Fig. 6-7. The molecular structure of allicin from garlic

ica], or *V. tahitensis* [from Oceania]). When dried it is long and black and imparts a delicious flavour to egg custard if the milk is boiled with a pod before being added to the beaten egg and sugar. Often for convenience cooks use vanilla essence. This is an alcohol (ethanol) extract of vanilla.

The chemical responsible for vanilla's exquisite flavour is vanillin. It is very soluble in alcohol hence the use of vanilla essence. Vanillin is not a natural component of the vanilla orchid, it is produced by the curing process – therefore it has no natural function in the plant.

Vanilla is expensive, and so often synthetic vanillin (either made from clove extract – see above, or synthesised chemically) is dissolved in alcohol as a cheap alternative. The main flavour chemical is exactly the same as in natural vanillin, but the other subtle flavours are missing.

Vanillin is of very low toxicity (LD50 [oral, rat]= 1,580 mg/kg; it would take about 100 g to kill a human).

Chilli

Chilli is a pepper (*Capsicum annuum*) of which there are many varieties ranging from the mildly flavoured sweet pepper that we use with onions, tomatoes and courgettes in ratatouille (a Provençal [S.E. France] vegetable stew), to the pungent birds eye chilli used in Indonesian and Thai cooking.

The burning sensation in your mouth after a curry, and the unspeakable burning at the other end of the alimentary canal the next morning are both due to the irritant chemicals in chilli. The most important is capsaicin which has an incredibly pungent taste and acrid vapour (you will know this if ever you have breathed in over frying chillis when making Thai food), humans can taste 1 part in 100,000 (i.e. approx. 0.001% solution) capsaicin.

Capsaicin is toxic (LD50 [oral, rat] = 100 mg/kg; about 6 g could kill a human), but you would have to eat a great many chillis to achieve this dose. Interestingly, its irritant properties are used by the police in immobilising sprays – just imagine how awful it would be to be sprayed in the eye with this highly irritant chemical.

It is clear that herbs and spices have numerous chemical constituents, most impart wonderful flavours and release mouth watering aromas when they vaporise during cooking. Some are incredibly irritant, others are toxic, but most are of relatively low toxicity. Some even have medicinal properties. The dose that you get eating food seasoned with herbs and spices is so small that toxicity is unimportant. Some people claim that they get benefit from the medicinal properties of herbs and herb extracts, but this is questionable because often the dose is often too low to result in a pharmacological effect.

When is a Food a Medicine?

[Reproduced in full, by permission of the Editor from Shaw, I.C. (2001) International Journal of Pharmaceutical Medicine 17:69]

There is an upsurge of interest in functional foods ("nutraceuticals") [*foods that have medicinal properties*], but is their use to replace real medicines justified? Before we can address this issue we must question why people appear more interested in these potential remedies than they did a decade or two ago. An idealist's answer might be that they cure ills without the adverse effects of conventional medicines and the need to visit the doctor. This hypothesis is almost certainly wrong! My cynical viewpoint is far more likely to explain their popularity, namely that advertising is making people think that functional foods work and that they are worth buying. All of this is just about to change as legislation around the world places anything with a medicinal claim firmly into the medicines camp. So if preparations of *Echinacea purpurea* (Purple Cone Flower from the American prairies) have the merest sniff of a claim to treat colds and influenza they will be considered medicines and not foods. At the moment in New Zealand there is a successful industry producing these, and other plants and their extracts. They are used to enhance the functionality of foods, or for export to other countries as components of herbal remedies which might be classified as either foods or medicines according to the regulations of the particular country.

Echinacea is a good example because it contains numerous phenolics that might well have pharmacological activity. Indeed, the purified phenolics have been shown to have effects in isolated cells and in *in vitro* systems. However, simple randomised placebo-controlled clinical trials have shown equivocal efficacy with respect to an enhanced phagocytosis endpoint; three of the five studies showed no efficacy (Melchart D, Linde K, Worku F et al (1995) Results of five randomized studies on the immunomodulatory activity of preparations of Echinacea. J Alt Comp Med 1:145–159). This might simply be a dose effect, but it points to the need for dose relationship efficacy studies with standardised test material.

Preparations of the herb are said to stimulate the immune system (possibly by enhancing phagocytosis) and so prevent cold, flu and minor infections. Yoghurt containing *Echinacea* is available in supermarkets in New Zealand (and elsewhere), it is claimed to "enhance the body's ability to resist infection" and to be "prized for its antibiotic properties". But are these claims true? The honest answer is that we don't know because they have not been tested in proper clinical trials, because *Echinacea* is regarded as a food and foods do not need to undergo clinical trials before they are marketed. But surely these are medicinal claims. The New Zealand Medicines Act 1981 defines a medicine as something having a "therapeutic purpose" and for use in "treating or preventing disease". In my opinion, the *Echinacea* yoghurt pot labels fall within the Medicines Act definition.

The world is waking up to this conundrum and Regulatory Authorities are grappling with the problems of medicinal claims and unconventional medicines. This will have a major impact on the nutraceutical and functional food industries. I suspect that either the foods (or are they medicines?) will disappear, or the medicinal claims will be removed from their labels because of the enormous cost of generating toxicological and efficacy data.

New Zealand has introduced an Advertisement Pre-Vetting Service run by the Advertising Standards Authority and Medsafe, through which all advertisements for "fringe" medicines must pass before the media will publish them. This is an excellent quality assurance system that helps to protect the consumer of these products. Despite this, advertisement vetting cannot replace toxicological and efficacy assessment. Surely the time is right for the manufacturers of ALL medicines, even if they prefer to call them foods, to have to support their claims with real science. And if they are not prepared to test their claims they should withdraw them.

Red Kidney Beans

Back in the early 1970s (when I was a student) there were several unexpected cases of severe illness amongst university students who created their version of the student all time favourite, chilli con carne, in crock pots (slow cookers) in their university lodgings.

These severe health effects were due to a group of potent protein toxins found in red kidney beans (*Phaseolus vulgaris*) – a crucial ingredient of "chilli". Red kidney beans contain a group of toxins called lectins. The two most important lectins in red kidney beans are arcelin and phasin. Phasin is a phytohemagglutinin – a plant chemical that causes blood to clot, and while present in other types of beans (broad beans have 5–10%

of the amount) its concentration is particularly high in the red kidney variety that we use for chilli con carne.

Phytohemagglutinins are large complex proteins that work by sticking red blood cells (erythrocytes) together by binding to one cell via its surface proteins and attaching another cell to another part of the protein molecule. This process leads to many cells being held together by the phytohaemagglutinin molecule (i.e. haemagglutination or clotting). This rapidly results in sever harm or death, because the clots block important small blood vessels (e.g. the brain's blood supply), and diminish the function of crucial organs.

Phasin is highly toxic, it takes only 5 µg/kg body weight to kill a human. This could be present in only one or two beans. So beware!

Does Cooking Beans Make Them Safe?

As is the case for most proteins, phasin is de-activated by heat. Therefore thorough cooking (i.e. at 100°C) significantly reduces the level of toxic protein and so makes the beans safer.

Levels of phasin in cooked and uncooked red kidney beans (data from Foodborne Pathogenic Microorganisms and Natural Toxins Handbook, US Food & Drug Admisistration at http://vm.cfsan.fda.gov/~mow/chap43.htlm)

Bean	Phasin level (phytohaemagglutinating units)
Uncooked	20,000–70,000
Thoroughly cooked	200–400

If cooked beans are safer, why did people get ill after eating crock pot-cooked chilli con carne? The answer is simple, food cooked in crock pots doesn't get to much above 60°C, and this is not hot enough to destroy the toxin. The solution is equally simple; make sure that red kidney beans are boiled for at least 10 minutes as part of their cooking process and all will be well. The alternative is to use canned beans that have been heated to a high temperature (121°) as part of the canning process.

Why Do Beans Contain Lectins?

It is often difficult to explain the purpose of toxins in plants. However, as discussed previously, pesticidal effects are often associated with natural plant toxins. Phasin and arcelin (particularly the latter) have insecticidal properties. This might explain why the plant expends so much energy to synthesise a complex protein molecule. But, perhaps the plant is simply trying to stop animals eating its seed, which after all is the plant's future.

Mycotoxins

Mycotoxins (from the Greek *Mukes* for mushroom [fungus]) are fungal toxins sometimes found in food that has been infected with a fungus. They are horrifically toxic and many are thought to cause cancer. The most important food mycotoxins are produced by fungi of the Genera Fusarium, Aspergilus and Penicillium. These fungi grow on carbohydrate-rich substrates like grains and nuts. They produce a range of mycotoxins, all of which are important food contaminants; most have specific legislation covering their levels in food. The most important mycotoxins and the fungi that produce them *Aflatoxins* are shown in the table.

Fungus	Mycotoxin
Fusarium graminearum	Deoxynivalenol, zearalenone
F. culmorum	Deoxynivalenol, zearalenone
Aspergillus flavus	Aflatoxin
F. verticilloides	Fumosin
Penicillium verrucosum	Ochratoxin

Aflatoxins

The aflatoxins (Aflatoxin-B1, -B2, -B2a, -M1, -M2, -G1, -G2 and -G2a) are perhaps the most notorious group of mycotoxin. They are a particular problem in peanuts which can become infected with *A. flavus* post-harvest. During storage and transport of the peanuts from the tropics where they are grown to temperate countries (e.g. USA) where they are very popular (e.g. as peanut butter), the fungus continues to grow and contaminates the nuts with highly toxic aflatoxins.

Very low doses of aflatoxin can have significant effects on consumers – Aflatoxin-B1 LD50[oral, rat] = 5 mg/kg body weight, which means that 300 mg could kill a human. Peanuts can contain 500 ug/kg if they are grown, stored and transported under conditions ideal for *A. flavus*'s growth. If in the exceptionally unlikely event that you ate only highly contaminated nuts, you would need about 600 kg of nuts to kill you. This, of course, is ridiculous, therefore the acute toxicity of aflatoxin is of little concern to consumers. Of very much greater concern is the long-term effect, namely cancer. Low doses over a long period of time might cause cancer. This is why regulators test peanuts and make sure that consumers are only exposed to low levels of aflatoxins that are unlikely to result in cancer. Codex Alimentarius (the international committee that sets standards for food) has set an MRL (maximum residues level) of 15 µg/kg for total aflatoxin in peanuts. This is a very low value that reflects concerns about these cancer-causing contaminants. It is illegal to sell peanuts with a level of total aflatoxins above the MRL.

Natural Toxins from Animals

Plants are not the only living things that produce toxic chemicals to ward off would be attackers. Animals also produce a broad array of ingenious natural toxins. Most of these (e.g. the snake venoms) are of no importance in the context of food (except perhaps to the pickers of tropical fruit who might encounter a poisonous snake). But there are just a few that could be included in food and present a significant risk to the consumer.

Tetrodotoxin and Sashimi

This intensely poisonous fish toxin has already been discussed in Chapter 2. It is produced by the Fugu Fish (a puffer fish) used for sashimi (raw fish). Tiny amounts can be fatal (see Chapter 2), but still Japanese people spend significant amounts of money on this prestigious delicacy.

Marine Toxins

These are not strictly speaking animal toxins, they are present in marine food animals (e.g. fish), but are derived from other creatures that have either been eaten by the animal, or have contaminated the animal after its death.

Histamine and Scromboid Fish Poisoning

An enzyme (histidine decarboxylase) present in the bacteria that colonise some warm water fish (e.g. Tuna – a *scromboid* fish) makes histamine from the amino acid histidine (present naturally in the fish). Histamine is intimately involved in allergic reactions (hence the use of anti-histamine drugs for allergy treatment), high doses of histamine cause symptoms similar to allergy. Histamine is present naturally in fish at levels of the order of 1 mg/kg, but in "toxic" fish concentrations can reach 100 mg/kg; the toxic threshold is 20–50 mg/kg.

Scromboid fish poisoning (or pseudo-allergic fish poisoning as it is sometimes called) is characterised by very rapid (as soon as 2 or 3 minutes after eating contaminated fish) onset of symptoms, including burning or swelling of the mouth, rash, diarrhoea, flushing, sweating, headache and vomiting. As you might expect treatment of severe cases is with anti-histamines.

The disease is not as rare as you might expect. In the USA there were 1,400 cases between 1973 and 1997, all were in states that have ready access to fresh fish.

Ciguatera Poisoning

This is associated with tropical reef fish (e.g. Red Snapper) consumption, and is caused by a toxin produced by a microscopic plant (a dinoflagellate – *Gambierdiscus toxicus*) that the fish eat. The toxin concentrates up the food chain, so large predatory fish (e.g. Barracuda) are more likely to harbour toxic levels of ciguatera toxin than their small prey.

Ciguatera toxin is an incredibly complex, large molecule that is horrifically toxic. It is amongst the most toxic chemicals known – LD50[i.p., mouse]=0.45 µg/kg body weight; 27 µg could kill a person and 0.1 µg can cause illness. The toxin works by interfering with the passage (transmission) of nerve impulses.

Ciguatera toxin's effects occur very rapidly after consumption of contaminated fish (as soon as a few minutes), and include a plethora of symptoms such as vomiting, diarrhoea, cramps, sweating, "pins and needles", and the strange sensation of reversed temperature sensitivity in the mouth – hot feels cold, and cold feels hot.

Ciguatera fish poisoning is common in parts of the world where reef fish are eaten. For example, in the US Virgin Islands and French West Indies it is estimated that 3% of the population are affected each year.

Paralytic Shellfish Poisoning (PSP)

This is caused by a different dinoflagellate (e.g. *Gonyaulax catenella*) to the species that causes ciguatera fish poisoning, and involves a different group of toxins. The PSP dinoflagellates are red and cause red tides when their numbers are so great (in excess of 50,000,000/ml) that they colour the water red. The Red Sea is so called because of its regular blooms of red algae.

The PSP dinoflagellates are filtered from sea water by bivalve molluscs (e.g. mussels) as a component of their microscopic diet. If you eat one of these shellfish you might suffer PSP. The symptoms include, tingling of the face, numbness, headache, weakness, partial paralysis, and rarely death. Symptoms occur as soon as 10 minutes (usually within 2 hours) after eating contaminated shellfish.

PSP toxins are a complex array of closely related molecules. There are likely to be many hundreds, all having similar toxic effects. They are often named after the shellfish from which they were originally isolated. For example, saxitoxin came from the Alaska butter clam (*Saxidomus giganteus*). PSP toxins are intensely poisonous; saxitoxin is the most toxic – LD_{50} [mouse, oral] = 263 µg/kg body weight. It would take only 15 mg of saxitoxin to kill a person. It works by interfering with nerve impulses, hence the tingling and numbness that it causes.

The Last Word

This has been just a brief foray into the world of natural toxins in food. I have covered a tiny fraction of the chemicals we eat each day with our food. I hope that it has persuaded you never to believe anyone who tries to convince you that natural is always safe! But also remember that even though you are eating these ingenious natural toxins every day, you are alive, and rarely succumb to their evil aspirations.

7
Agrochemical Residues in Food

7   Agrochemical Residues in Food

Most people will tell you that their greatest worries about food are residues of chemicals from farming (agrochemical residues) that food might contain, and genetic modification (see Chapter 9). This perception of the risk is likely to be wrong. Toxicity studies suggest that the levels of agrochemical residues that are usually found in food are far below those that will harm the consumer. There is still concern in some scientist's minds (including mine) that we don't understand fully the effects of long-term exposure to residues, or worse, the long-term exposure to cocktails of residues. Regulators might try to convince us that residues are safe, or have negligible risk, but they really don't know because the studies necessary to make these statements simply have not been carried out. Despite this, the evidence that we have suggests that the risk is low.

The fact that consumers regard chemical residues as the biggest problem is a good example of risk perception (see Chapter 2). In reality the risk is likely to be much lower than the perceived risk. This is certainly the case for exposure to single chemicals.

What Are Agrochemical Residues?

Agrochemicals are chemicals used in farming, including:

- Pesticides – herbicides, insecticides, fungicides
- Veterinary medicines
- Fertilisers (e.g. nitrate)

When these chemicals are used they are applied to crops and animals that will eventually become our food. If, for example, a fungicide is sprayed onto wheat to control rust (a fungal disease characterised by an orange/brown powdery residue on the plant that looks like iron rust), the fungicide will first fall onto the plant. Its concentration on the outside of the plant will be very high – it would likely be very risky to eat the plant at this stage. As time passes, a proportion of the fungicide might be taken up by the plant, while some might be washed off by rain, and some degraded by air and UV light. The fungicide remaining in and on the plant is termed the *fungicide residue*. The same applies to other pesticides, vet medicines and fertiliser components.

Back to the fungicide scenario; if the wheat in our example is harvested and milled into flour, the fungicide residues might then be present in the flour. If they are on the husk of the wheat, they are likely not to contaminate white flour because the husk is removed, but will contaminate whole-wheat (brown) flour because the husk is incorporated into the flour. If the flour is made into bread, the residues will be in the bread. Then when you eat a nice whole-meal bread ham sandwich, you will get a dose of the fungicide with your lunch.

The concentration of fungicide in the final product (e.g. bread) will be millions of times less than the concentration on the wheat immediately after the farmer sprays his crop. The reason for this is in part due to the decrease in residue concentration due to degradation (e.g. by UV light effects), dilution (not all of the wheat milled into flour will have residues), and by the effects of cooking.

Many agrochemicals are chemically unstable, so if they are heated they break down. Residues in raw foods might decline considerably on cooking or processing. In our flour scenario the bread is cooked at 200°C or more, this is likely to break down some agrochemical residues. Work in my own lab has shown that the organophosphorus pesticide (OP), triazophos decreases when apples containing residues are cooked. Important questions are: what are the breakdown products? Are they more or less toxic than the parent agrochemical residues? We don't have the answers to these questions for most residues but there is evidence emerging that some agrochemicals are degraded to potentially toxic chemicals (e.g. pyrethroid insecticides), while with others, the toxicity is lower. But more of this later.

Regulations Protect us from Residues

Many of the chemicals that farmers use to grow our food are very toxic. They are designed to kill fungi, bacteria, nematodes, insects, weeds, and a myriad other organisms that contrive to reduce the farmer's productivity and therefore profitability. Because of their toxicity agrochemicals are highly regulated. The regulations are intended to minimise both farmers' and consumers' exposures and the possible harm that might result.

Standards are set to minimise exposure and make trade in foods with unacceptable residues illegal. These safeguards are an excellent way of protecting the consumer providing that farmers and traders comply with them. Compliance is assured in most countries by regular surveillance of food to make sure that residues are below the levels set in the standard. The process of approving an agrochemical (e.g. a pesticide) for use in all developed countries involves a consideration of the acceptability of residues that will result in food if the chemical is used properly. To most people these safeguards are adequate but some people simply do not want agrochemical residues in their food – whether regulators and scientists say that they pose a miniscule (if any) risk or not. These people tend to favour organic food because agrochemical residues should be lower than in conventionally produced food. This, of course, is their choice, and their right... but don't forget natural toxins... and the increasing evidence that some of them might be present in higher concentrations in organic food (see later in this Chapter).

Agrochemical Regulation

All developed countries have government-driven processes for approving agrochemicals for use (i.e. licensing). The systems are similar, I'll use the UK's pesticide approval system as an example. The Pesticide Safety Directorate (PSD – an agency of the government's Department for the Environment, Food and Rural Affairs [DEFRA]) is responsible for approving pesticides. It operates via an independent advisory committee – the Advisory Committee on Pesticides (ACP), which has members taken from the great and good of the pesticides world. There are experts in human toxicology, agronomy, risk assessment, etc. This is an excellent (I might be biased though – I was a member until I moved to New Zealand) committee that very seriously considers a new pesticide from every angle before making a decision. Most importantly, the Committee is independent and is not influenced by government or industry. The PSD and DEFRA have no say in the decision whether or not to approve, they merely provide the administration to act on the ACP's deliberations. Some countries (e.g. New Zealand) don't have an independent advisory committee, but rely upon a government department to make approval decisions.

The ACP reviews the pesticide on the basis of its efficacy and safety.

Efficacy – does it work? Is it useful? $\rightarrow$ **BENEFIT**
Safety – Is it toxic to consumers, $\rightarrow$ **RISK**
farmers and the environment.

They make their decision on a risk:benefit basis. It is acknowledged that we must accept risk, but that it must be outweighed by the benefit of the (in this case) pesticide. The risk side of the equation requires that the company applying for approval of a new pesticide, new formulation, or new use carry out significant experimental work in compliance with strict guidelines to assess toxicity (e.g. extensive animal toxicity studies, for both long- and short-term effects). On the benefit side of the equation is the usefulness of the pesticide, is it better (e.g. less toxic) than currently used pesticides?

If approval is granted, it will come with conditions. For example in order to minimise the consumer's exposure to residues there will be a specified period between application and harvest (based on farm studies submitted as part of the approvals package by the agrochemicals company). This will be measured against standards (e.g. an acceptable residue level).

Once a pesticide has been approved for use, the manufacturer can market it in line with the approved claims and crop/animal combinations (e.g. it might have approved use on wheat for control of rust). This means that it cannot be used for any other purpose without further regulatory approval. Once it is in use it is important to make certain that the approved use is the only use, and that the use regimen (e.g. withholding time) is obeyed.

Regulatory Standards – MRL

The MRL (Maximum Residue Level) is a regulatory standard; it is the concentration of an agrochemical in food that might result from the proper use of the chemical in agriculture. Proper use is defined in a set of regulations called Good Agricultural Practice (GAP). The MRL is not a measure of toxicity, but rather a trading standard. Despite this MRLs are scrutinised by toxicologists to make sure that if you eat a food containing residues at or below the MRL you should come to no harm. MRLs are agreed internationally via Codex Alimentarius.

Basically if a food contains a residue below the MRL it can be sold, and is unlikely to cause the consumer any ill effects.

Withholding Time

This is the time taken between application (or dosing if it's a medicine) of the agrochemical to reach the MRL. It is set by Regulators as a condition of approval. If it is obeyed it is very unlikely indeed that an MRL will be exceeded, and therefore that a consumer will suffer ill effects.

Surveillance

Most developed countries carry out extensive surveillance schemes to police the use of agrochemicals. Continuing our UK pesticide scenario, the Brits have an independent committee (the Pesticide Residues Committee – PRC; formerly the Working Party on Pesticide Residues – WPPR) that co-ordinates and deliberates on surveillance for pesticide residues in food. They commission work to investigate pesticide levels in foods (e.g. OP residues in bread), and decide whether standards are being breached. If so, they attempt to trace back the food to its producer and take appropriate action – this can be severe (i.e. large fines). The PRC includes toxicologists and analytical experts and has government money (about £2,000,000 in 1999) to fund its surveillance programme – its independence and expertise makes it a good consumer protection committee (again, I might be biased – I chaired the Committee 1995–2001). Surveillance is an essential part of the regulatory armoury for ensuring that residues do not pose a significant risk to consumers. Countries without surveillance schemes cannot, in my opinion, assure their public of its food's safety.

How Do we Know How Toxic an Agrochemical is?

The toxicology studies used to determine safety as part of the approval's process can be used to determine two important parameters, the No Observable Adverse Effect Level (NOAEL), and the Acceptable Daily Intake (ADI). These are real measures of toxicity.

NOAEL

This is sometimes called the No Effect Level (NEL) or No Observable Effect Level (NOEL); it is the highest dose of a chemical (in this case an agrochemical) that causes no signs of toxicity.

ADI

This is the dose of a chemical if given every day for a lifetime that will result in no measurable pharmacological (or toxicological) effect. It is calculated from the NOAEL:

$$ADI = \frac{NOAEL}{SF}$$

where SF is the safety factor (usually 1,000).

This is rather extreme. Safety of agrochemicals is based on receiving a dose of the chemical every day for your entire life – this is extremely unlikely, so we base our safety assessment on an extreme worst case. I think that this is the right thing to do because it enables me, as a toxicologist, to sleep at night! And it should give consumers confidence in their food.

It is important to view toxicological measures (ADI) and standards (MRL) in context. MRLs are sometimes exceeded (in

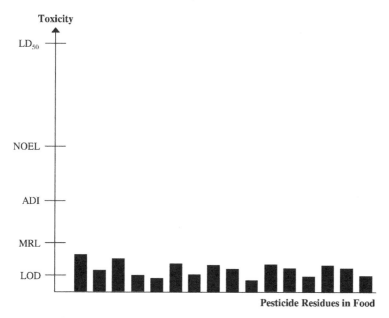

Fig. 7-1. The toxicological hierarchy of LD50, MRL, ADI, NOAEL, analytical limit of determination (LOD) and pesticide residues in food

the UK, about 1% of food surveyed contains residues at or above MRLs), while ADIs are only exceptionally rarely exceeded (in my 6-years chairing the WPPR I only saw one near ADI exceedance). Exceedance of an ADI required immediate action, e.g. withdrawal of the food from the market. You might think this a little extreme when you consider that an ADI exceedance relates to a single analytical result in a single sample, but that the definition of an ADI relates to exposure to that level for an entire lifetime. This is the "better safe than sorry" approach to life – or more scientifically, application of the precautionary principle (Fig. 7-1).

Pesticide Residues in Food

Consumers get worried about pesticides in food, but as discussed earlier this concern is probably greater than it should be on toxicological grounds. I am happy that the international regulations protect us well. If we look at surveillance for residues around Europe you can see that there are very few MRL exceedances, and don't forget the MRL is way below the level at which toxicologists get worried.

Pesticide MRL exceedances in Europe in 1996 has been adapted from Shaw IC (1999). Pesticides in Food. In: Brooks and Roberts (eds) Pesticide Chemistry and Bioscience – the Food Environment Challenge. RSC Press, London, pp 421–428

Country	% of Samples above MRL	Country	% of Samples above MRL
Belgium	1	Luxembourg	1
Denmark	1	Netherlands	0
Germany	0	Portugal	1
Greece	1	Finland	3
Spain	0.1	Sweden	2
Ireland	3	Norway	0
Italy	1	UK	<1

The rest of the developed world has similar values. The developing world is more of a problem, because there is often little pesticide legislation in operation – they have bigger things to think about, like feeding malnourished and starving populations. In this context, who cares about pesticide residues? Consumers of third world produce do! Northern countries import a vast amount of developing world-produced fruit and vegetables (e.g. bananas from South America). The importing countries take account of the lack of certainty about residues in such produce by imposing import regulations and residues surveillance. Some food outlets (e.g. supermarket chains) go even further by contracting producers in warmer countries to farm, using procedures (e.g. pesticide use) acceptable to their home market. This is a win-win situation. The farmer gets a lucrative contract with certainty of sale of his produce, and the food outlet gets produce that its customers want.

MRLs are the tip of the residues iceberg. We should also consider residues below the MRL. But before we do we must get the ground rules straight. It is difficult to compare residues from country to country and year to year because the laboratory methods used to detect chemicals in food have improved over the years – we are detecting residues now that we could not find 10 years ago, therefore if we look over time it might appear that the residues situation is getting worse. But in fact it is probably the chemists that are getting better. Comparing between countries is difficult too because they use different methods with different detection limits. For this reason I will restrict myself to one country (UK) and cover a time period that used comparable analytical methods with comparable detection limits (Fig. 7-2).

You can see that MRL exceedances in the UK are below 1% and about 35% of food had measurable pesticide residues over a considerable period. So the situation appears to be static. In fact this probably represents the residues that result from modern farming practice; we must live with them if we accept modern farming. The only way that we will lower residues is to change the way we farm.

There is an important worldwide movement that advocates turning the farming clock back to the time when there

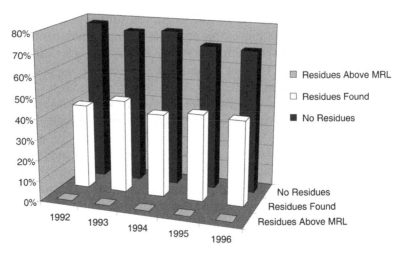

Fig. 7-2. Pesticide residues below the MRL, and MRL exceedances in food in the UK (*data from MAFF [now DEFRA], UK*)

were no OCs, OPs and pyrethroids, no Superphosphate and ammonia-based fertilisers mined from far off lands, and no chemicals that make our farm animals grow faster and bigger. This is the time when the sun shone all summer, and a tweed clad little boy ran across the fields of skipping lambs to take his dad a whole meal cheese sandwich for lunch. This was a time of nostalgia and happiness. This is the organic movement – but more of organic production later in this chapter.

A Near ADI Exceedance in the UK – Lindane in Milk

Lindane is a controversial OC pesticide that is either banned or use restricted in most developed countries. It has been linked to breast cancer by some lobby groups; the evidence is scant, but this and its persistence in the environment have made it a hot topic of debate amongst toxicologists and environmentalists alike. It is approved in the UK for a few very specific uses, e.g. in the early stages of sugar beet growing.

So, lindane is a controversial pesticide that many people think should be banned. You can imagine the furore that raged

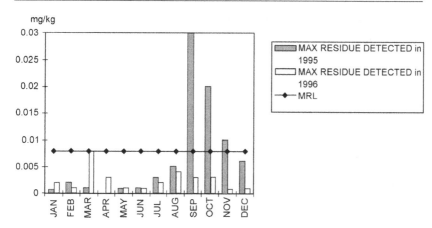

Fig. 7-3. Lindane levels in milk in 1995/96 in the UK showing a near exceedance of the ADI. There was very great concern about this, rightly so, but in toxicological terms it would only have been dangerous if levels had remained above the ADI for someone's lifetime... this is not likely ... no, it's impossible! (Data from the Annual Report of the Working Party on Pesticide Residues 1996, p 21, MAFF Publications, London)

when it was found in milk in the UK in June 1995. In subsequent months' samples, lindane residues continued to rise, reaching a peak in September (Fig. 7-3). The milk levels in September were only marginally below the ADI. Residues near to the ADI in a staple dietary commodity (i.e. milk, potatoes or bread) are very worrying. Fortunately October's residues were significantly lower and therefore a potential crisis did not come to fruition. It was important to try to explain the effect in order to try to prevent its reoccurrence. In this case there were two main contributory factors:

- The summer of 1995 was a hot summer in the UK – it is possible that milking cows were marginally malnourished and so to maintain milk output they might have released fat reserves. It is well known that animal fat harbours long lived lipid-soluble residues (e.g. lindane), therefore lindane from fat reserves might have been mobilised and incorporated into milk.
- The summer's drought had resulted in a poor cereal, grain and forage crops, therefore these animal feed components were imported. The imports might have contained lindane residues, and were incorporated into cattle feed.

These possible explanations could not be proved, or even investigated, but in my opinion they are both likely to have contributed to the effect.

In 1996, lindane residues in milk began to rise again. This happened at approximately the same time as the previous year's enormous increase. Regulators held their collective breaths – would 1995's problem recur? I was almost afraid to look at the lindane surveillance results as they came from the lab, but thankfully they made a feeble attempt to rise above baseline levels. We could all breathe again! It is possible that the second small rise was due to farmers feeding their cattle left over feed from the previous year – so the lindane dose was much lower than in 1995. By 1998 there were no measurable residues of lindane in milk.

Isofenphos in Pigs

In the UK in 1990, a strange "disease" appeared in pigs in the north of England. Thousands of pigs began to stagger around, dragging their back legs. There was a ripple of terror that this might be a manifestation of BSE in pigs, this was amongst the UK government's worst fears at the time. Investigation by my Group at the Central Veterinary Laboratory showed that the cause of the "disease" was poisoning by a rare OP approved for use in France but not in the UK. The pesticide is called isofenphos; it was used to dress (i.e. coat) seed grain to kill insects that might fancy the planted grain for their lunch. Isofenphos is very toxic (LD50 [oral, rat]=28 mg/kg body weight), causing blockage of nerve impulses and inhibiting a specific central nervous system enzyme called neurotoxic esterase (NTE) which if blocked results in ataxia (i.e. dragging of the legs, inability to walk straight) – the UK prohibits the use of all NTE inhibitors. Analysis of fat from affected pigs revealed very low concentrations of isofenphos. So the symptoms and toxic agent agreed, but where on earth had isofenphos come from?

Further detective work traced the source of the isofenphos to a French warehouse – by now the story was beginning to sound like a cheap 1940s detective film. Isofenphos-dressed grain was stored in the warehouse alongside another grain product called wheat screenings (i.e. husks and other unwanted byproducts of the grain industry), and wheat screenings are used in animal feed manufacture. So we had a connection. Indeed, when we picked through the wheat screenings we found tell-tale orange specks that had originated from the orange dyed (as a warning) dressed grain. This saga illustrates perfectly the need to control every stage of food production in order to protect consumers from residues – this is a classic food chain contamination. The following table shows the route of contamination of pigs in the UK with isofenphos in 1990.

The route of contamination of pigs in the UK with isofenphos in 1990 (data from Shaw IC et al. (1995) Veterinary Record, 136, 95–97)	
Sample	Isofenphos concentration
Dust from French warehouse floor	18.3 mg/kg
Wheat screenings	156 mg/kg
Pig feed	1.8 mg/kg
Pig fat	0.01 mg/kg

The next question was, are the pigs safe to eat? On toxicological grounds the answer was a resounding, yes! The highest level found in fat (isofenphos is fat soluble, so this is where the highest residue levels would be found) from affected pigs was only 0.01 mg/kg which is way below the level of toxicological concern to humans. Despite this, the affected pigs were not used for human consumption.

Post-Harvest Pesticides

There are two main uses for pesticides, to control pests during a crop's growth period, or during storage. It is perhaps the latter group that represent the greatest risk to consumers because they are added to the harvested product in store and therefore have less chance to degrade before we eat them.

A typical example is pirimiphos-methyl, an OP used on stored grain. Such pesticides are essential if bulk grain is to be stored, since the storage silos soon become infected with insects (e.g. weevils) that rapidly grow and eat their way through the valuable product. Pesticides such as pirimiphos-methyl put paid to this, but at a cost. They persist because there is no UV light and only low microbiological activity in the storage silo to break them down, and therefore they form residues in the grain which later becomes bread. A quick look at residues of pesticides in bread makes the point and the following table shows pirimiphos-methyl found in bread and wheat in the UK in 1999.

Pirimiphos-methyl found in bread and wheat in the UK in 1999 (data from the Annual Report of the Working Party on Pesticide Residues (2000), Ministry of Agriculture, Fisheries and Food, London)

	Number of samples analysed	% Positive	Residue range (mg/kg)
Wheat	62	13	0.05–0.1
Brown bread	58	12	0.06–0.3
White bread	84	2	0.06–0.07

There is more pirimiphos-methyl in brown bread because brown flour is made by grinding the whole grain which includes the husk. When the pirimiphos-methyl is applied to the stored grain its highest concentration is on the outside of the grain. In fact for this reason there is a remarkable agreement between residues statistics for wheat and brown bread. When the husk is removed to make white flour a good proportion of the pirimiphos-methyl residue goes with it which explains why residues in white bread are lower.

What Effect Might Pirimiphos-Methyl in Food Have on our Health?

Pirimiphos-methyl's LD50[rat, oral]=140 mg/kg body weight, therefore it would take about 8 g to kill a person. Clearly it would not be possible to eat enough bread in a single sitting to achieve this dose – in fact, based on the highest residues in brown bread, you would need to eat 27 tonnes of bread to be sure of dying from pirimiphos-methyl poisoning! Death is the upper end-point of acute toxicity, so let's explore how much pirimiphos-methyl you would need to consume to suffer a pharmacological effect, such as salivation (this is a typical OP toxicity due to stimulation of nervous impulses to the salivary glands, causing them to over produce saliva). Pirimiphos-methyl's NOAEL[rat, oral]= 0.5 mg/kg body weight/day (dosed every day for 80 weeks), this means that it would take at least 30 mg in a single dose to have an effect on a person – this equates to 100 kg of bread with the highest residue at a single sitting. Therefore, acute toxicity is not an issue.

Very much less is known about the effects of multiple doses over long periods of time, but remember that the rat study from which the NOAEL was derived involved daily doses to the rats each day for 80 weeks. I am happy that the small amount of pirimiphos-methyl in my food is safe. Even if you add up all of the pesticides that we eat each day, I'm still not worried. What I *am* worried about though is the environment. When pesticides are used, huge concentrations enter the immediate spray environment. The effects on animals and plants living there is great. Concentrations diminish rapidly with distance from the sprayed field, but still we can see effects on wildlife. I find environmental impact a far more compelling argument to reduce pesticide use than human effects via residues in food. Similarly, the Organic Movement appeals to me ecologically, but not necessarily in a food safety context.

Excipients

Excipients are the chemicals added to pesticides as part of the commercial formulation. They help distribute the active ingredient (i.e. the pesticide itself) and enhance its absorption by the pest, and so maximise its pesticidal effect. They are often detergents, and slowly it is emerging that some have hitherto unexpected toxicity profiles all of their own. For example nonylphenol (a non-ionic detergent) is used in some insecticide formulations; it is an estrogen mimic (see Chapter 8) and might be more ecotoxic, and toxic to consumers than was originally thought. Therefore when assessing the impact of pesticides on people and the environment we must not forget the excipients.

Measuring Human Exposure to Pesticides

It is not ethical to give people doses of pesticides to see what doses are toxic, although accidental poisonings and exposures provide useful information in this respect. Therefore, food safety scientists use surveillance and total diet surveys (TDS) as tools for assessing intake, They then compare intake with toxicity measures (e.g. ADI) to assess risk to the consumer.

Surveillance and Dietary Surveys

Surveillance studies measure residues in food, then we use dietary surveys to determine how much of a particular food we eat (at a population level). By multiplying the two together it gives us an idea of the intake of a particular chemical. Comparing this intake with NOAELs and ADIs we can estimate risk to the consumer.

Total Diet Surveys (TDS)

TDSs provide a snapshot of dietary intakes and assess intake in relation to what is actually eaten. In a TDS food is prepared in the way that it normally would before being eaten – bananas are peeled, steak is fried, potatoes are boiled and mashed, flour and dried fruit are baked in cakes, etc., therefore processing effects are taken account when pesticides are determined in the food that we eat (as discussed earlier in this chapter, some pesticides are degraded during cooking).

A very good TDS is carried out by the New Zealand Food Safety Authority and Institute of Environmental Science and

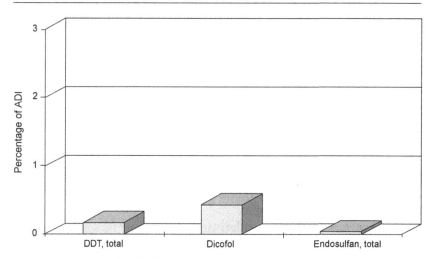

Fig. 7-4. Organochlorines in the New Zealand diet as a percentage of the ADI, showing that the total amount is less than 1% (data from Cressey et al (2000), 1997/98 New Zealand Total Diet Survey, Part 1 Pesticides, Ministry of Health, Wellington)

Research [ESR] (I might be biased here, I live in New Zealand [NZ], headed (2000–2004) the ESR Programme that carries out the TDS, and sit on the New Zealand Food Safety Advisory Board). The last NZTDS (1999) addressed OC intake in terms of ADI. Its findings were very interesting (Fig. 7-4).

The total intake of OCs by New Zealanders in 1997/98 was less than 1% of the ADI. Therefore it is very unlikely indeed that a NZ consumer will come to any harm due to OCs in their food. Intake calculations can be made for other developed countries, they are remarkably similar.

Pesticides in Human Fat

It is very difficult to assess accurately how much pesticide gets into our bodies. We can calculate how much we eat from the levels in food and the amount of food we eat, but this does not take account of the percent of the pesticide in our food that is absorbed into the body. It is very likely (supported by animal studies) that the uptake (i.e. the amount we absorb into our bodies) is lower, in some cases very much lower, than pesticide intakes (i.e. the amount of pesticide eaten with food).

In addition to the difference between intake and uptake, we must also consider metabolism and excretion. Some chemicals are absorbed into the body, but are rapidly metabolised to less toxic forms and excreted in the urine or bile. Pesticides that are absorbed efficiently, then rapidly metabolized and excreted are likely to have a lesser impact on the body than those that hang around for a long time. Most of the modern pesticides (e.g. OPs) are more efficiently metabolized and excreted than the old-style pesticides (e.g. OCs) and therefore are likely to have less of an impact on our bodies.

Many pesticides (particularly insecticides) are fat soluble, therefore measuring their levels in human body fat is a good way of assessing uptake and body burden. There are two types of fat that have been used to assess human exposure to pesticides:

1. Body fat – taken by biopsy from living people, or post mortem
2. Milk fat – taken during lactation.

Pesticides in Human Body Fat

The UK's Working Party on Pesticides Residues reported some interesting studies on human fat samples collected between 1995–1997. All of the 203 samples analysed contained OCs, some at quite high concentrations. This is perhaps not surprising because the fat sampled came from people who died "naturally" and so had lived through the hey day (i.e. the 1960s) of OC use. During their life times they would have accumulated OCs from their food, water intake and other environmental and occupational exposures. In the following table showing OC pesticides in human fat in the UK.

Perhaps the most interesting finding is that DDT breakdown products were found in 99% of the samples. Therefore most people have been exposed to DDT and have residues in their bodies. Whether these residues are doing any damage is very difficult to decide. I suspect that they are quietly sequestered

OC pesticides in human fat in the UK (data from the UK Working Party on Pesticide Residues Report, MAFF, 1996)

Pesticide	Concentration range (mg/kg)	Number of samples in range
Chlordane**	Not found	95
	0.01–0.1	108
DDT**	Not found	2
	0.01–0.09	19
	0.1–0.9	135
	1.0–9.3	47
Dieldrin	Not found	83
	0.01–0.1	120
β-HCH*	Not found	3
	0.01–0.09	99
	0.1–0.8	101
γ-HCH*	Not found	197
	0.01	5
	1.9	1
Heptachlor**	Not found	142
	0.01–0.05	61
Hexachlorobenzene	Not found	13
	0.01–0.09	175
	0.020.1–0.2	15

* Different chemical forms of the insecticide Lindane (hexachlorocyclohexane – HCH).
** Measured as metabolites/breakdown products.

in fat and are completely harmless – it might even be that fat is the body's means of holding on to potentially harmful fat soluble chemicals. The problem, of course arises when fat is mobilised during times of nutritional stress (e.g. lactation; see below).

Pesticides in Human Milk

Human milk is particularly interesting as a means of assessing exposure to pesticides because levels in first lactation milk in theory represent body burden (i.e. accumulation of pesticides) from birth. Levels in second lactation milk represent exposure since first lactation because lactation clears residues (and passes them on to the suckling infant!). This is useful because it gives an indication of exposure at different times in the subject's life.

There have been some good studies from around the world on pesticides in human milk. They show beautifully how legislation affects people's exposure to pesticides. In countries with good legislation (e.g. USA) residues are much lower than in countries with lax regulation (e.g. India). This is illustrated particularly well by Germany before unification; women from the former East Germany (Deutsche Demokratische Republik – DDR) have much higher milk pesticide levels than their West German counterparts. It is likely that East Germany had less stringent pesticide use regulations than their near neighbours. The women were separated only by regulations, they were geographically very

DDT in human milk from around the world – the German results show that political separation and different legislation affects people's exposure to pesticides (data from Shaw et al. (2000) Environmental Science and Pollution Research, 7, 75–77)

Country/city	DDT (p,p'-DDT in mg/kg)
USA	
Arkansas	0.039
New York	0.023
Germany*	
E. Berlin	2.28
W. Berlin	0.81
Thailand	
Bangkok	0.734
Vietnam	
S. Vietnam	4.22–7.3
Ho Chi Minh	0.023
Papua New Guinea	0.42
India	
Ludhiana	7.18
Faridkot	13.81

* Data collected before German unification.

close and therefore the differences cannot be explained by different agricultural needs (e.g. due to climate differences).

What do the milk pesticide residues tell us? Firstly the only pesticides found are the long-lived OCs. This fits in well with the human body fat work. So it is very likely that other pesticides are metabolised and excreted rapidly and therefore will have little long-term impact on the body. Also, the levels are low, and if this is an accurate assessment of exposure over many years – we really are not exposed to much. Indeed, several of the OCs found in different studies are now banned (e.g. DDT) and therefore the residues are probably due to past use. So if we carry out similar studies in a decade's time it is likely that residues will be lower.

Pesticides excreted in milk might be a good way for scientists to assess exposure, but milk was not "designed" as a tool for scientists' exposure assessments, but rather as nutrition for infants. So what will pesticide in milk do to the infant? The only honest answer is, we don't know. However there have been many attempts to assess intakes and risk. On balance, it is likely that levels are usually low enough not to pose a significant risk, especially since the suckling period is quite short and therefore represents a relatively small total intake. I am more concerned that mothers' worries about pesticides in their milk might lead them not to suckle their kids. By not feeding naturally they lose the enormous benefits (including transfer of immunity from mother to child) of their own milk. In addition, replacement milks (e.g. soy or cow's) also have associated risks; soy is rich in phytoestrogens (see Chapter 8), and cow's milk is likely to contain lower concentrations of a similar spectrum of pesticides to mother's milk. On balance, mother's own milk is best!

Vet Medicines

Farm animals get ill just like we do, so they need medication. The problem is that the medication might hang around in their bodies so that when they are slaughtered we might get a small dose of the medicine in our meat. There are strictly applied withdrawal periods for pesticides to minimise the consumer's exposure.

The risk from veterinary medicine residues in our food is very much lower than the tiny risk from pesticides because medicines are designed to have low toxicity, whereas pesticides are designed to be toxic.

There are two types of vet medicines, those used on the outside of animals (e.g. to kill parasites – parasiticides) and those used internally. Some medicines for external use are pesticides – the OP Diazinon is used in sheep dips to kill lice and mites, and as a spray against insect pests. So we might get residues of these chemicals in our food from two different sources.

Most developed countries run veterinary residues monitoring programmes to police the proper use of vet medicines and minimise human exposure via food. The UK's Veterinary Residues Committee (VRC) in 2001 reported over 7,000 analyses for veterinary medicines, of these 155 (i.e. about 2%) exceeded MRLs.

Antibiotic Residues, Growth Promotion, and Antibiotic Resistance

Medicines can be used either to cure or prevent disease. Vets can prescribe medicines for mass medication of herds or flocks if they fear that they might be at risk from a disease. For example, if a respiratory disease has affected a pig herd a vet might medicate a neighbour's herd as a safeguard. Antibiotics are sometimes used in this way.

Some antibiotics cause animals to grow faster, either by directly affecting the animals' metabolism, or by preventing low-grade infections and making them feel healthier and hungrier. This is useful to farmers who struggle to compete in the vigorous livestock market. For this reason it is possible that farmers might use antibiotics (and other agents such as hormones) to promote growth instead of in the treatment of disease. The use of growth promoting agents is frowned upon in most developed countries except the USA where they are still used extensively.

The problem with antibiotic use is the development of resistant bacterial strains. These are bacteria that develop biochemical mechanisms to reduce their susceptibility to the toxic effects of antibiotics. Resistant strains are unaffected by antibiotics. It is therefore impossible to treat resistant infections with the antibiotic to which they are resistant. Unfortunately bacteria can develop resistance to more than one antibiotic – multiple resistance. Bacteria can pass their resistance factors (fragments of nucleic acid that code for resistance) on to another bacterium. This is a significant problem because if the widespread use of antibiotics in farming (particularly the mass use as growth promoters) leads to resistance in bacteria in farm animals, these bacteria might find their way into humans and exchange resistance factors with natural gut bacteria. Then if the person harbouring the resistant strain in their gut gets infected by a pathogenic bacterium, it is possible that the pathogen might pick up the resistance factor and be untreatable with antibiotics. In serious cases this might result in death of the infected person.

Antimicrobial resistance is a very real problem that has led many countries to seriously review the use of antibiotics, and in particular ban, or advise against, the use of antibiotic growth promoters.

From the vet residues point of view, banning growth promoting antibiotics is a good move. This will most certainly reduce residues levels in meat.

Have Vet Medicines Caused Harm to Consumers?

As discussed above vet medicines are designed to be safe to animals, therefore it is unlikely that they will cause harm to people below the therapeutic dose. It would be almost impossible to get a therapeutic dose from residues in a piece of meat. However people who are particularly sensitive might be harmed, or even killed, by vet medicine residues in meat. For example, meat residues of penicillin to some one with a penicillin allergy could have very serious consequences.

Clenbuterol

There are very few cases of vet medicines having harmed people – i.e. the risk is tiny. One case, however, stands out. This in- _____

volved the respiratory drug clenbuterol. Clenbuterol (a β-agonist) is used both in animal and human medicine in the treatment of respiratory disorders such as asthma. It dilates the airways making breathing easier. It is a good drug with few side effects if used properly. Like other β-agonists it can cause problems in people with heart disease.

If used properly clenbuterol residues are rarely, if ever, found. However, farmers discovered that at high doses it aided their animals' growth. In fact it did more than just make the animals grow, it changed the proportion of fat to lean meat in favour of lean. This had the potential to significantly increase the value of the carcass. Clenbuterol therefore began to be used illegally in growth promotion (as a repartitioning agent, i.e. made the animals produce lean rather than fatty meat).

The illicit use of clenbuterol came to the fore between 1989 and 1990 in Spain when 135 people were poisoned by clenbuterol residues in meat. The drug accumulates in liver (where it is metabolised) therefore the human poisoning cases were all associated with eating liver. There have been several other outbreaks of clenbuterol poisoning reported in Spain, France and Canada. A farmer even died in Ireland preparing laced feed for his animals – rough justice! The dose of clenbuterol necessary to have a pharmacological effect in people is very small – 5 μg is enough to bronchiodilate and speed the heart rate in most people. Levels in liver were 160–500 μg/kg in the Spanish residues cases, so 10 g of liver might be enough to have a pharmacological effect on the consumer – most people would eat about 100 g at a sitting which brings them well into the toxic range for clenbuterol.

Clenbuterol is, perhaps the only example of a vet medicine that has harmed consumers of meat. There are other vet medicines that might have caused harm, but data do not exist to support disease incidence associated with residues consumption. For example, chloramphenicol, an antibiotic now banned for use in animals intended for human consumption, causes a rare form of leaukaemia (aplastic anaemia). There is no evidence that cases of the disease resulted from residues in meat...but who knows?

Are Vet Medicine Residues a Problem?

Vet medicines are the least worrying of all residues, but they still cause disproportionate angst amongst activists. The worry is far greater than is justified by the problem, but it is important to keep Regulators on their toes to maintain good surveillance schemes to police their countries' farmers' use of these important animal welfare aids. Shame on those countries that do not have vet medicines surveillance schemes even if the risk is low!

Fertilisers

As farming becomes more and more intensive to meet the demands of the world market and the desires of the farmers for greater profits more and more fertiliser is needed to allow the land to support the enormous productivity demanded of it. Millions of tons of fertilisers are used worldwide annually. In New Zealand (a small country with a large agricultural export market) alone, 2,800,000 tonnes of fertiliser was used in 1998. These fertilisers have a huge environmental impact because they increase nutrient levels in waterways and lakes resulting in excessive growth of algae (algal blooms) and other plants (e.g. reeds) resulting in huge demands from these fragile aquatic environments. Eventually the demand from the plant growth outstrips their environment and the algae die and rot, and bacteria colonise and over-stretch the oxygen provision of the water and the system becomes unbalanced and eventually collapses (this process is termed eutrophication).

This enormous use of fertilisers also results in high levels of some of the fertiliser components in food plants. For example, nitrate levels are becoming a problem in some leafy vegetables such as lettuce. A recent (2002) study in Italy showed that the highest dietary sources of nitrate were chicory and rocket, and that lettuce contained much lower levels, but because it comprised a significant part of the Italian diet it could account for as much as 60% of the dietary intake of nitrate. Adding all of these (and other) intakes together still does not exceed the ADI (remember this is the amount of a chemical that can be consumed every day for an entire lifetime without harm).

Nitrate levels in salad vegetables (data from [1] DeMartin & Restani (2002) Food Additives and Contaminants 20, 787–792, and [2] Ysart et al. (1999) Food Additives & Contaminants 16, 301–306)

Salad vegetable	Nitrate level (mg/kg)
Rocket	6,25[1]
Chicory	6,12[1]
Lettuce	3,30[1]
Spinach	1,90[2]

The table lists the nitrate levels in salad vegetables.

Interestingly organically grown food often contains more nitrate than its "conventional" counterpart. So organic food eaters are at greater risk from nitrate, as are vegetarians who tend to eat more leafy greens.

But why all the fuss about nitrate intake? Nitrate (NO_3^-) can be chemically reduced in the gut to nitrite (NO_2^-) which might react with other dietary components (secondary amines) to produce highly carcinogenic nitrosamines which cause gut cancer in animals – and are very likely to do the same in humans. Fertiliser-derived nitrate/nitrite is not the only dietary source of nitrate/nitrite. Cured meats (e.g. bacon) contain very high levels indeed because potassium nitrate is used in the preservation process.

Since we all eat nitrite/nitrate-containing food you might ask, why don't we all get gut cancer? One explanation for this is that other chemicals in our diet (e.g. ascorbic acid – vitamin C) inhibit the chemical reaction that forms nitrosamines. Many leafy vegetables have high levels of vitamin C and so even though they contain nitrate the nitrosamine forming reaction is inhibited. This is a good illustration that we should consider diet as a whole when assessing potential adverse effects, rather than focusing on one nasty component of one food in a very complex diet.

Heavy Metals

Heavy metal is a chemical term which describes metallic elements with high atomic weight (e.g. lead). Teenagers have a quite different definition – there is even a heavy metal band called LD50 (how appropriate!). Back to the real heavy metals. Many are highly toxic, including cadmium (Cd), lead (Pb), and mercury (Hg).

Heavy metal compound	LD50 [rat, oral] (mg/kg body weight)	Main toxic effects
CdCl2	88	Acute – death Chronic – cancer
Pb(NO$_3$)$_2$	93	Acute – death Chronic – nerve damage/anaemia/possibly cancer
HgCl$_2$	1	Acute – death Chronic – nerve damage
NaCl	3,750	Acute – *Very* high doses can be fatal

The Pb(NO$_3$)$_2$ values were not available for PbCl$_2$ in the rat, therefore LD50 [mouse, ip] given. The toxicity of common salt (sodium chloride – NaCl) is shown for comparison.

Cadmium (Cd)

Some foods can have relatively high levels of heavy metals and present an unacceptable risk to consumers. A good example is shellfish (e.g. mussels) which filter vast volumes of water to extract the microscopic plants and animals that they feed on. If the water, or microscopic creatures, contain heavy metals the metals will be concentrated by the shellfish. They are concentrated because the shellfish have proteins that bind some heavy metals. If you eat a meal of contaminated shellfish it is possible that you will get an unacceptably high dose of the heavy metal. New Zealand is a volcanic country, and cadmium is a metal associated with volcanic activity. Therefore the silts in New Zealand's coastal waters can contain high levels of cadmium. Shellfish inhabiting these areas can have correspondingly high levels of cadmium. It is for this reason that the New Zealand Ministry of Health recommended that shellfish not be eaten more than once a week. New Zealand is not the only country with this problem. Other volcanic countries will be the same, as will countries which pollute their marine environments with heavy metal-containing industrial effluent.

Cadmium levels in New Zealand food (data from Cressey (p 72))	
Food in New Zealand	Cadmium level (mg/kg)
Bacon	0.003
Egg	0.002
Lamb's liver	0.113
Oysters	4.481
Carrots	0.027
Cabbage	0.005
Lettuce	0.021

Note the Cd level in oysters, it is extremely high, data from Cressey et al. (2000), 1997/98 New Zealand Total Diet Survey, Part 1 Pesticides. Ministry of Health, Wellington, p 72

The ADI for cadmium is 1 µg/kg body weight (i.e. about 60 mg/person/day). Human dietary exposure is about 35 µg/day of which 2 µg (6%) are absorbed (i.e. it is poorly absorbed). There is a high level of cadmium in cigarette smoke, so smokers have a considerably greater intake than non-smokers (another reason to give up!). Shellfish can have 4,000 µg/kg cadmium, so you would only have to eat 2 g of shellfish (i.e. part of an oyster) with this level to exceed the ADI.

Cadmium in Kidney

Offals, particularly kidney, accumulate cadmium, so people who eat a lot of kidney might be exposed to higher levels of cadmium. This is particularly so if the kidney is from an older animal. Horse kidney is the greatest problem because if eaten it is generally derived from older animals. The UK Veterinary Residues Committee recently (2001) reported a study in which 100% of 8 samples of horse kidney analysed exceeded the Cd MRL. Similarly 13 of 27 (48%) deer kidney samples analysed in the same study exceeded the MRL; one had a Cd level of 16,400 µg/kg (MRL=1,000 µg/kg). Clearly it is not a good idea to eat deer or horse kidney. The table shows values from horses in Italy.

Cadmium levels in horse tissues from Italy (data from Balzan et al. (2002) http://www.lnl.infn.it/~annrep/readAN/2002/contrib_2002/B011_B117T.pdf)

Tissue	Cadmium level range
Kidney	47–1192
Liver	38–92
Muscle	73

Why is Cadmium so Toxic?

Cadmium ions (Cd^{2+}) behave in the body like calcium ions (Ca^{2+}). Calcium is extremely important in many cellular biochemical functions, cadmium inhibits these reactions and kills the cell. It also interferes with the uptake of calcium by the body; calcium is absorbed on a specific carrier protein that binds the calcium and carries it through the membrane that surrounds cells, cadmium binds to the same membrane carrier protein so preventing calcium uptake. For this reason long-term exposure to cadmium is associated with bone weakening (osteomalacia).

Cadmium is one of the few metals that can cause cancer after long-term exposure. However the doses needed to cause cancer are high for prolonged periods of time – this is highly unlikely from food.

Mercury (Hg)

Mercury can occur in three forms, organic (the correct use of the word, as opposed to "organic" food which is a popular misnomer), inorganic, and elemental (mercury metal – not important in a food context). Organic mercury includes a carbon-containing moiety in its molecule (e.g. methyl mercury – $Hg(CH_3)_2$)) which makes it very fat soluble and therefore absorbed into the body well – for this reason it is exceptionally toxic. Inorganic mercury is a simple mercury salt such as mercuric chloride ($HgCl_2$) and generally is fairly water soluble (high water solubility = low fat solubility) and therefore absorbed into the body less efficiently than organic mercury, and therefore is far less toxic.

Toxicity of organic and inorganic mercury compared with their water solubilities (high water solubility = low fat solubility)

Mercury compound	Water solubility	Toxicity – LD50 [rat, oral]
Organic Methyl mercury – $Hg(CH_3)_2$	Insoluble	Very high, possibly ng/kg
Inorganic Mercuric chloride – $HgCl_2$	480 g/l	1 mg/kg

Minamata Mercury Poisoning – a Classic Mercury Poisoning Case

Minamata is a Japanese fishing town on Kyushu (Japan's southernmost island). In the early 1930s, a company was set up to manufacture acetaldehyde for use in the plastics industry. This brought much needed employment to the area and was heralded as a great advance. Mercury was used in the acetaldehyde manufacturing process and the factory's mercury-containing effluent was pumped out into Minamata Bay. In the early 1950s fish deaths were noted in the bay, and pet cats began to develop bizarre behavioural changes. Soon afterwards some of the local fishermen and their families became "crazed", convulsed, and became extremely ill with characteristic extreme salivation. Many people were affected and some died. In 1956 mercury poisoning was identified as the cause of the ills.

This is a classic case that involved inorganic mercury effluent from the factory being released into the marine environment where bacteria in the silt converted it to organic mercury (mainly dimethylmercury). The oily fish absorbed the fat soluble inorganic mercury, some died if the dose was high enough, others were caught by the fishermen. The highest mercury levels were in the fish fat (e.g. brain which is fat rich) – the fishermen fed their cats the fish heads which is why the cats became ill first. The fish flesh contained lower levels of mercury, but mercury accumulates in the human body so with each meal the fishermen's families accumulated more mercury, until they reached toxic levels and became ill themselves.

In 1970, the company responsible was ordered to pay US$3.2million to clean up Minamata Bay and so make fishing safe once more.

Is Mercury Still a Problem in Food?

Mercury levels can be quite high in long-lived oily fish because they pick up mercury from sea water and accumulate it in their bodies. Eating such fish can deliver undesirably high mercury intake, but is unlikely to result in toxicity to the consumer unless fish is a very regular (i.e. most days) component of their diet. However the bits of the fish that we don't usually eat (e.g. fish heads) might be used by the poultry feed industry – it is obvious where this tale is leading – so the mercury contamination is concentrated and transferred to chicken. There is now concern about this route of mercury exposure and some countries monitor poultry and poultry feed for mercury to ensure that unacceptable levels are not finding their way into our Sunday roast chicken or coq au vin.

Why is Mercury so Toxic?

Mercury binds to sulphydryl (–SH) groups which are very important in many cellular reactions. If mercury is bound, the biomolecule becomes inactive (its molecular structure can be completely changed by mercury) and therefore cannot perform its vital cellular functions. Mercury targets nerve cells where –SH groups are important in the process of neurotransmission – mercury is a neurotoxin. Organic mercury is much more toxic than inorganic mercury, partly because it is absorbed from food better, and partly because once in the body it is readily taken up by nerves and the brain (neurological tissue is fatty) so being concentrated in its target action site.

The symptoms of mercury poisoning manifest after prolonged exposure; they include tremors, stumbling gait (ataxia), numbness and tingling (parasthesia) particularly around the mouth, difficulty in speaking, and many other neurological symptoms. Death is a common endpoint. The phrase "as mad as a hatter" comes from the use of mercuric chloride in wool felting. The felt was contaminated with mercury which meant that hat makers were exposed to high doses of mercury and showed the signs of toxicity. – Lewis Carol's Mad Hatter is a good example.

Lead (Pb)

Lead is the most abundant heavy metal in the Earth's crust (14 mg/kg compared to Cd – 0.1 mg/kg; Hg – 0.05 mg/kg) and so is more likely to find its way into our food "naturally". Like mercury it exists as organic and inorganic forms, the former being far more toxic for the same reasons as discussed for mercury. Inorganic lead, however, is not converted to organic lead by living organisms – organic lead (tetraethyl lead) was extensively used as an anti-knock additive to petrol, but was withdrawn in most countries when its toxicity was realised. Most of the lead in our food came from lead from petrol settling out of the air onto food plants. Its levels declined rapidly after lead in petrol was banned in the developed world.

Lead exposure from food in adults is about 150 µg/day of which about 15 µg (10%) is absorbed – to add further grist to my anti-smoking mill, smokers absorb a further 6 µg/day from cigarette smoke. The US Food and Drug Administration (FDA) have set a tolerable daily intake (TDI – i.e. "safe" intake) for lead of 75 µg/day for adults. It would be difficult to reach this level if you have a varied diet.

Essential Dietary Metals

It seems strange in a book about food safety to discuss essential elements, but if they are absent from our diet this makes our diet unhealthy, or perhaps even unsafe.

Selenium (Se)

Se is essential to human life, but at tiny intakes – a normal healthy diet provides about 100–200 µg/day. It is important in biochemistry as an antioxidant – it mops up reactive chemicals produced metabolically that might be harmful to cells. If it is eliminated from the diet of experimental animals they suffer all sorts of ills, including congestive heart failure and cancer. There are some well described deficiency diseases in humans, for example Keshan Disease – congestive heart failure in kids living in parts of China where Se is deficient in soil.

Se is present in most diets in most parts of the world at sufficient levels to provide for consumers' daily needs. However there are some places where Se is not at sufficiently high levels in the soil to maintain sufficient levels in crops to provide for human dietary needs. Parts of China (see Keshan Disease above) and New Zealand are good examples.

Most of our intake of Se comes from grains, particularly wheat (via bread). In countries where Se levels in soils are too low to provide for consumers' needs all that needs to be done to redress the balance is to import a proportion of the country's wheat from a selenium-rich country. New Zealand does this – wheat is imported from Canada to boost Se intakes. The other way would be to fortify flour with Se, but this is likely to be controversial. For some reason people will accept naturally incurred chemicals, but frown upon addition of the same chemical to their food, even if it is good for them!

Iodine (I)

Iodine is important because it is a constituent of the thyroid hormones (secreted by the thyroid gland in the neck) which are crucial for growth, development, and metabolism. If your thyroid hormones are low (i.e. hypothyroidism) you will be lethargic, and if this occurs early in life growth and development will be impaired (cretinism). Low levels of thyroid hormones in adults is characterised by goitre, this is a swelling of the thyroid gland in the neck – the gland swells in an attempt to trap more iodine from the blood in order to make more thyroid hormone (Fig. 7-5).

Iodine is often low in soils of volcanic countries, and therefore goitre is more prevalent in people from these countries. South East Asia and New Zealand are good examples of iodine deficient regions.

The New Zealand total diet survey illustrates iodine deficiency very well indeed. In fact it shows clearly that iodine intake is declining. This is due to a decline in salt consumption (sea salt contains iodine, table salt is fortified with iodine) because of fears about the health effects (e.g. coronary heart disease) of salt that have been supported by government health campaigns, and, strangely, the decline in use of iodine-containing disinfectants in the dairy industry – the very low iodine contamination of milk as a result of their use contributed towards New Zealand's iodine intake. New Zealand has to do something to increase iodine intake otherwise thyroid deficiency and goitres will re-appear (Fig. 7-6). The following table shows iodine levels in milk in New Zealand compared to the UK.

Iodine levels in milk in New Zealand compared to the UK (data from New Zealand Total Diet Survey 1997–98, NZ Ministry of Health)

UK	0.2–0.4 mg/kg
New Zealand	0.06–0.2 mg/kg

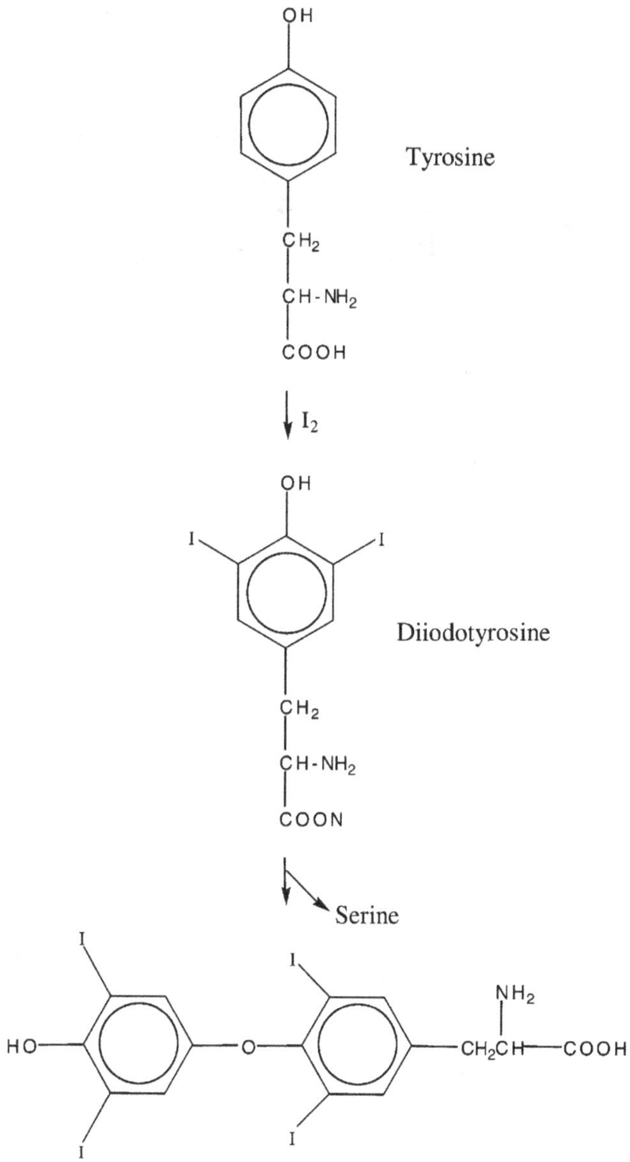

Thyroxine (T_4)

Fig. 7-5. Synthesis of thyroxin (T4) from tyrosine and iodine in the thyroid gland

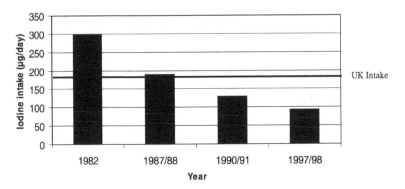

Fig. 7-6. Trends in dietary iodine intake for young males in New Zealand compared with the dietary intake in the UK (data from New Zealand Total Diet Survey 1997–98)

There are many more examples of both nasty and essential metals in our food, far too many to cover here.

Organic Food

To chemists *organic* means containing the element carbon as part of the molecule, to biologists it means pertaining to life, however more recently a movement has developed that advocates the production of food without the use of artificial chemicals (except for a handful of approved substances) and often using more traditional farming methods that impact little on the environment. This is a laudable approach to farming – anything that reduces environmental impact, and so protects our fragile environment, is good. In addition, some people think that organic food is better for you, is free of nasty pesticides, and tastes better. But are their beliefs founded?

There are few good scientifically robust studies that have compared organic and conventionally-produced food and so it is difficult to definitively determine the facts about organic versus conventionally-produced food. Despite this I'll look at three key factors as a means of shedding some light onto the debate.

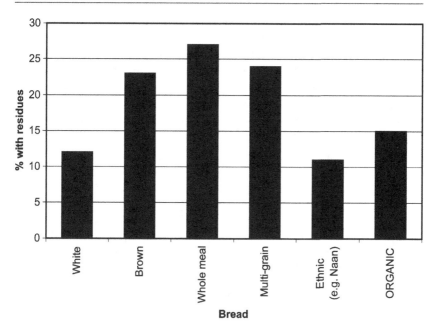

Fig. 7-7. Pesticide residues in organic compared with conventionally-produced breads sold in the UK between 1988 and 1996 (data from UK Working Party on Pesticide Residues (1996), MAFF, UK)

Are There More Pesticide Residues in Organic Food?

Since synthetic pesticides are not used in organic farming we would expect residues in food to be low. We might not expect them to be zero because there is a general background contamination of land with pesticides because of their long-term and continued use by conventional farmers. So if you buy organic food labelled *pesticide free* beware, it probably isn't! All that organic farmers can guarantee is that they have not used pesticides themselves. This is illustrated well if you look at pesticide residues in organic versus conventionally-produced bread sold in the UK (Fig. 7-7).

Are There More Bacteria on Organic Food?

Organic farmers use more "natural" fertilisers (e.g. farmyard manure) than their conventional counterparts. Manure comes from animals' bottoms and is laden with bacteria. Fertilisers often used by conventional farmers are derived from mined minerals (e.g. Superphosphate) and therefore are ostensibly bacteria-free. They both deliver essential nutrients (e.g. nitrogen as nitrate) to crops and so do not have a great effect on such things (e.g. nitrate, see earlier in this Chapter) in the food derived from the crops. So you might think that organic food is more likely to have bacteria lurking on its surface than conventionally-produced food. But several good studies have shown that this is not the case and that there is no difference in pathogenic bacterial (e.g. *E. coli* 0157) contamination of organic versus conventionally-produced crops. But why is this so? It is likely that organic farmers understand traditional production methods and therefore compost their farmyard manure well before applying it to their crops; the composting process kills many pathogenic bacteria so making the "fertiliser" safe.

Are Natural Toxins Higher in Organic Food

Many natural toxins present in fruits and vegetables have pesticidal properties (e.g. insecticidal), and are used by the plant to prevent attack by pests (e.g. insects) – see Chapter 6. Therefore if a farmer does not use pesticides on his crops the crop might produce its own defence against pest attack. It is therefore possible that organic crops might contain higher levels of natural toxins than conventionally-produced crops.

Nutrients in Organic Food

Organic officianados will tell you that organic food contains more nutrients than its conventional counterpart. I had always pooh poohed this until I looked into the scientific literature ... and I found that they are right. The differences in levels of 3 important food "nutrients" between organic and conventionally-produced crops are shown in the following table.

Differences in levels of 3 important food "nutrients" between organic and conventionally-produced crops (data from Firman Bear Report, Rutgers University, USA)

Crop	Calcium (mEq/100 g)	Thiamin (mEq/100 g)	Iron (mEq/100 g)
Cabbage			
Organic	60	13	94
Conventional	17.5	2	20
Lettuce			
Organic	71	169	516
Conventional	16	1	9
Tomatoes			
Organic	23	68	1938
Conventional	4.4	1	1
Spinach			
Organic	96	117	1584
Conventional	47.5	1	19

It is likely that nutrient levels are higher in organic crops because the crops take longer to grow and so accumulate more nutrients than their forced conventional counterparts.

Is Organic Food Better for You?

This is the $64,000 question. A review of the scant data comparing organic and conventional produce suggests that pesticide levels are lower in organic food, bacterial contamination is the same, natural toxins might be higher in organic food, and some nutrients are higher too. Pesticides are neither here nor there, because levels in conventional food pose a negligible health risk; bacterial contamination is the same, some of the natural toxins can be very nasty (e.g. psoralen a carcinogen – see Chapter 6) and might be higher in organic food; some nutrients are higher in organic produce, but you probably would get enough in a well balanced conventionally produced diet anyway... it's equivocal. There's one thing for certain, organic production is very much better for the environment, it is for this reason that I would choose organic food.

Take Home Message

Agrochemical residues are not of enormous concern from the point of view of food safety – although we need to keep an eye on research results around long term exposure to cocktails of pesticides before we sit back and relax. One thing is for certain, reducing our use of pesticides and fertilisers would be better for our fragile environment; my concern about agrochemicals is environmental not food safety-related.

8
Gender Bending Chemicals in Food

8   Gender Bending Chemicals in Food

In 1996 a very interesting, and rather unexpected, scientific paper was published by Louis Guillette, a scientist working at the
University of Florida, USA. The paper reported a reduction in
the penis size of alligators living in Lake Apopka in the Florida
Everglades. The effect was blamed on contamination of the lake
by agrochemicals, in particular DDT. These chemicals were
shown to mimic the female hormone (17β-estradiol) and result
in feminisation of creatures exposed to them. These estrogen
mimics were termed xenoestrogens (*xenos* – from the Greek
for foreign). Soon after Guillette's work was published there
were numerous reports of similar effects in a multitude of different species, including reproductive disorders in French oysters, and female Dogwhelks (a marine gastropod) growing
penises in the UK.

Four years earlier a study had been published in the
British Medical Journal which discussed the steady decline in
human sperm count over the past 50-years. There was much scientific interest in these two observations because it just might
be that the same chemicals that caused the effects in wildlife
were responsible for the declining sperm count in men. This
seemed a bit far-fetched at the time and there were many articles in learned scientific journals proposing connections, and
then equally well-founded papers refuting the evidence. Science
works on this toing and froing of opinion, eventually agreement is reached. In the case of xenoestrogens the science jury is
coming firmly down on the side of these chemicals having
significant effects on the reproductive organs of wildlife *and*

humans. The most likely routes of exposure for humans are food and water.

What Are Estrogens? How Do They Work?

Estrogens are the female hormones. They are responsible for the sexual (and some other) characteristics of females, including:

- Growth of breasts
- Growth of female sex organs
- Influence mood
- Bone density – postmenopausal women sometimes suffer from weak bones (osteoporosis) because estrogen levels decline when their periods stop
- Slow down skin aging.

Estrogen works by binding to a specific receptor protein in cells. When bound the protein changes its shape and moves into the cell nucleus and switches on the genes that cause feminisation. The interaction with the estrogen receptor (ER) is like a lock and key – estrogen is the key to the receptor's lock (Fig. 8-1).

Estrogens comprise several different molecules, all with the same basic molecular backbone, the most important for our discussion is 17β-estradiol – this is the most potent feminising hormone. The hormone responsible for women!

Both males and females have 17β-estradiol, and both males and females have ERs in their cells. There is a spectrum of maleness and femaleness; someone with high estrogen levels and a large number of ERs will be at the very female end of the sexuality spectrum. Someone with very few receptors and very low estrogen levels will be at the tip of maleness. It is not quite as simple as this though, there are other hormones involved (e.g. the male hormone testosterone), and it is the ratio between the levels of male and female hormones, combined with the receptor numbers that determines the position on the sexual spectrum. It is fascinating that metabolically 17β-estradiol is de-

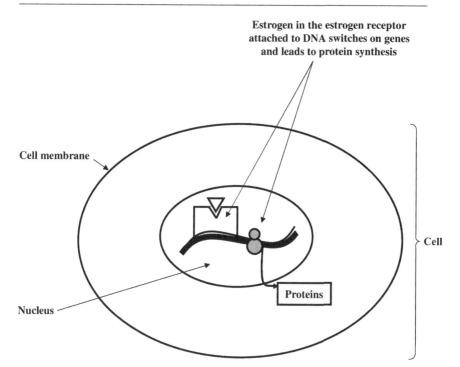

Estrogen in the estrogen receptor attached to DNA switches on genes and leads to protein synthesis

Cell membrane

Cell

Proteins

Nucleus

Fig. 8-1. Estrogen's mechanism of action, showing how 17β-estradiol (estrogen) via its interaction with the estrogen receptor (ER) inside a cell switches on genes that code for proteins that result in feminisation

rived from testosterone by just 3 reaction steps, and that the hormone series is derived from cholesterol... so cholesterol is not all bad. Even the blokeyest bloke is just 3 metabolic steps away from being a woman! (Fig. 8-2).

Other Chemical Keys Fit the ER Lock

There are specific parts of the 17β-estradiol molecule that interact with specific sections of the ER. For example the two –OH (hydroxyl) groups on the estrogen molecule are known to interact specifically with amino acids in the receptor site. Also the long water repelling (hydrophobic) region between the hydrox-

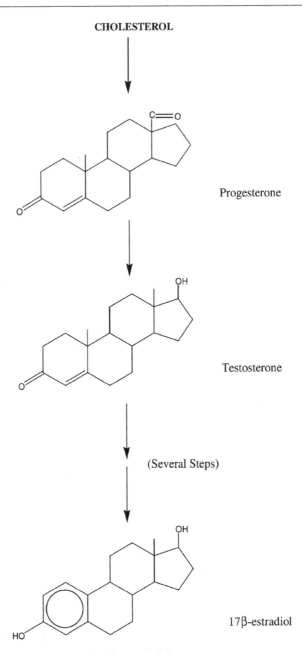

Fig. 8-2. The formation of 17β-estradiol from testosterone – men are just a few enzymes away from being women!

yl groups interacts with a sequence of water repelling amino acids in the ER. These facets of interaction on the 17β-estradiol molecule are akin to the notches on a latch key; if they interact with the lock they will activate the lock's mechanism and open the door. The ER is just the same, providing the molecular key has the right notches to activate the mechanism cellular feminisation will result.

There are many molecules that have structural analogies to 17β-estradiol and therefore fit the ER. Some of them are a sufficiently good fit to activate the receptor and initiate feminisation. Perhaps the best example is diethylstilboestrol (DES). This was used as a fertility drug in the past – this is hardly surprising since the ER is a key player in fertility, ovulation, and conception. Later DES was used as a growth promoting agent (see Chapter 7) in meat production because it stimulates fattening (as does 17β-estradiol). A quick look at the molecular structure of DES shows very clearly why it fits and activates the ER (Fig. 8-3).

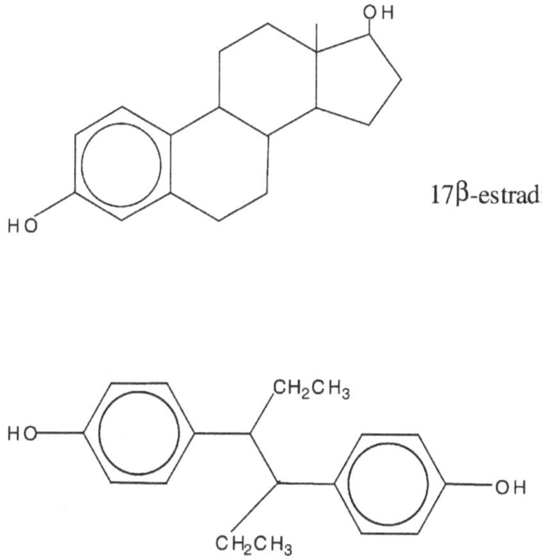

17β-estradiol

diethylstilbestrol

Fig. 8-3. Molecular structures of DES and 17β-estradiol showing how similar they are – it is obvious why DES fits the estrogen receptor and results in feminisation

DES residues are no longer an issue in food because most countries have banned its use because of an association with cancer. However, there are many other estrogen mimics (xenoestrogens) that are present in food, and many scientists (including myself) believe that they are contributing to the strange effects (e.g. reduced sperm count) occurring in men.

Xenoestrogens in Food

These chemicals fall into two classes:

- natural constituents
- contaminants.

Contaminants are further divided into environmental pollutants that find their way into food by food plant and animal uptake from the environment, and chemicals (often plasticizers) that originate from food packaging and manufacturing processes.

Natural Xenoestrogens in Food

Plants contain a vast array of chemicals, which are presumably important in the plant's life processes. Some help the plant to fix energy from the sun (by the process of photosynthesis), others are used to store energy (e.g. starch in potatoes), some prevent the deleterious effects of reactive chemicals produced by normal metabolic processes (e.g. flavonoids) and others act as natural insecticides and fungicides to protect plants from pests (see Chapter 7). Some of these natural chemicals are xenoestrogens – particularly the flavonoids and the natural pesticides (many are phytoestrogens). Therefore there are xenoestrogens present naturally in many vegetables ... and always have been.

I will use genistein, a natural plant xenoestrogen as an example to illustrate the point. Genistein is an isoflavone phytoestrogen found in soy (and other beans) (Fig. 8-4).

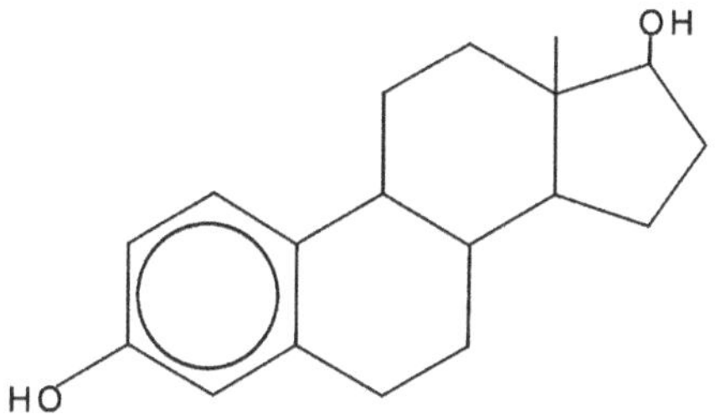

17β-estradiol

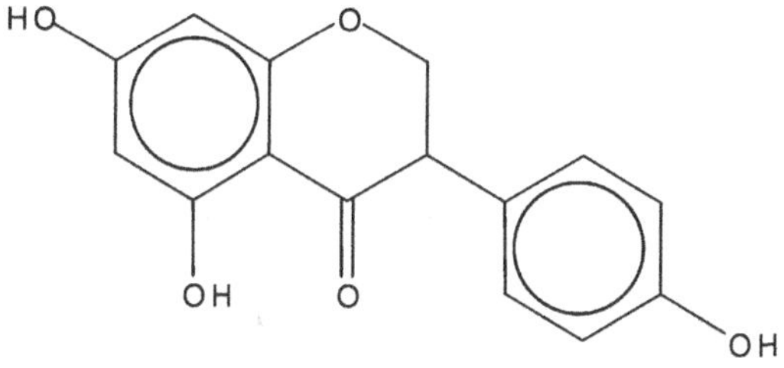

Genistein

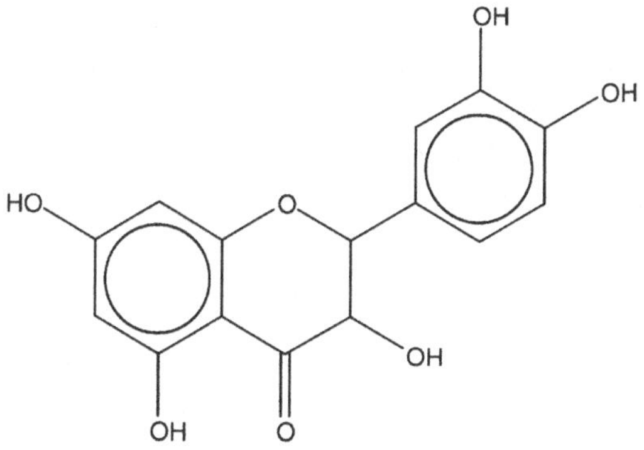

Quercetin

Fig. 8-4. Molecular structures of genistein (from soy) and quercetin (from oranges) showing how closely they resemble 17β-estradiol – it is obvious why they are female hormone mimics

Ingestion of 60 g of soy protein per month (equivalent to 0.7 mg isoflavone/kg body weight/day) has been shown to suppress hormones (e.g. leutenising hormone [LH]) at the peak of the menstrual cycle. This is just one source of xenoestrogens, the diet contains many more, therefore the likelihood of a pharmacological effect is great – but more of this later.

Plant xenoestrogen	Source
Daidzein	Chickpea
Quercetin	Citrus fruit (e.g. oranges)
Kaempferol	Grapefruit
Phloretin	Apples
Genistein	Soy, beans, spinach

Just a few of the xenoestrogens that might be in your veges.

Synthetic Xenoestrogens in Food

The Lake Apopka alligator story began concerns about synthetic xenoestrogens in the environment. It is thought that the short alligator's penises were due to exposure to xenoestrogen environmental pollutants such as DDT and nonylphenol (an industrial detergent). Just like the natural plant xenoestrogens fit the ER so do many synthetic estrogen look-alikes. Sometimes the molecular similarity is not as obvious for these chemicals, but once you understand some of the principles of chemistry it is easier to see. For example, hydroxyl groups attract electrons to themselves so becoming slightly negatively charged (the chemists call this electronegativity, the –OH is said to be δ- [i.e. a little bit negatively charged]), this effect is due to the oxygen atom's desire to acquire more electrons. Other atoms do the same – i.e. they are electronegative. Chlorine is electronegative, and becomes δ- in much the same way as –OH. The interaction with the ER actually needs a δ-chemical group not necessarily -OH which explains why DDT, with two chlorine atoms at opposite ends of its molecule, is a xenoestrogen (Fig. 8-5).

This brief chemistry lesson is important if we are to understand the structure activity relationships (SARs) between the synthetic xenoestrogens and 17β-estradiol.

As mentioned above there are two main groups of synthetic xenoestrogens:

1. Environmental pollutants, e.g. pesticides, industrial chemicals
2. Chemicals from food packaging and manufacturing processes, e.g. plasticizers.

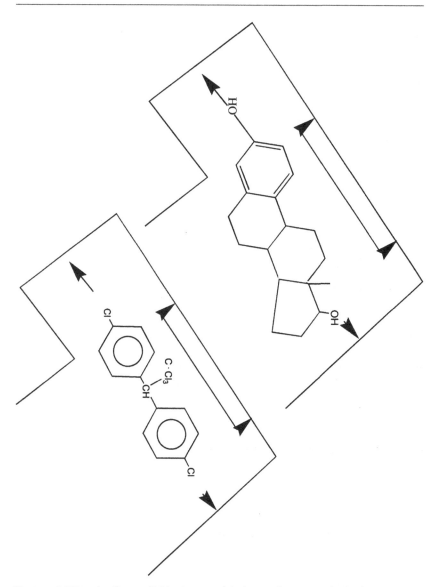

Fig. 8-5. DDT and 17β-estradiol in the ER – this shows why DDT unlocks the receptor's feminising activity

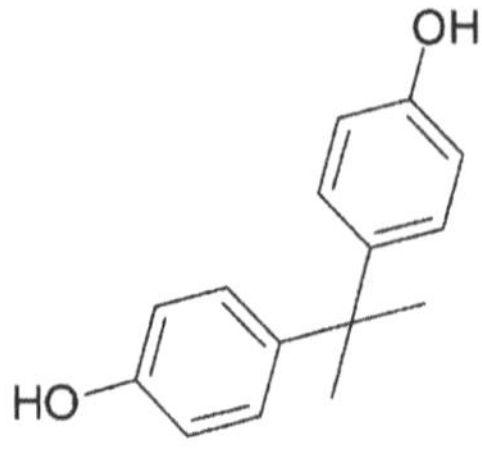

Fig. 8-6. Bisphenol-A and 17β-estradiol – the similarities are clear

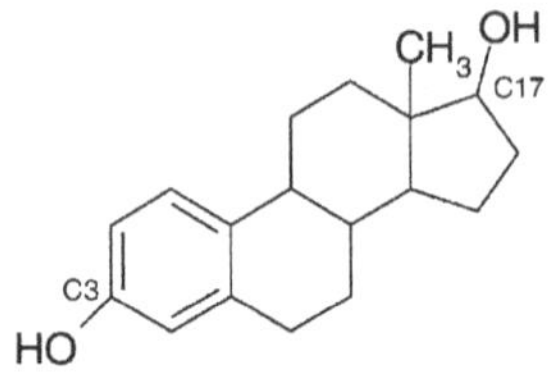

DDT and nonylphenol fall into category 1. Bisphenol-A, a chemical used in the manufacture of lacquers sometimes used to coat the inside of food cans, falls into category 2. There are many more examples of xenoestrogens in these classes (Fig. 8-6).

Do Xenoestrogens in Food Affect Humans?

There is much controversy about this, however there is a steadily increasing body of evidence that some of the effects that we are seeing in humans might be caused by xenoestrogens. It is easy to show what effects xenoestrogens have in animal studies, and the effects are those that we are currently seeing in the human population (e.g. declining sperm count), and it is easy to show how much xenoestrogen we are exposed to, but it is impossible to produce a definitive cause effect relationship without giving men doses of xenoestrogens over a long period of time and observing the effects – this would both take a long time and not be ethical (I can't imagine many male volunteers coming forward to take part in this experiment!).

My research group has been studying intakes of xenoe-strogens and trying to determine what effect these might have. We have calculated the amount of specific xenoestrogens that we ingest, modelled the levels that these represent in blood, and compared these with normal blood levels of 17β-estradiol. We relate xenoestrogen level to the corresponding level of 17β-estradiol necessary to give the same effect (i.e. relative estro-genicity). If the xenoestrogen relative estrogenicity level is high-er than the normal 17β-estradiol blood level a pharmacological effect is possible. There are numerous arguments against this approach because some important assumptions are made as part of the model (e.g. that xenoestrogens are 100% absorbed from food – this is very unlikely). However it is a start and al-lows us to look at the relative importance of the different xe-noestrogen classes: Theoretical blood levels of xenoestrogens from food in the UK are shown in the next table (data from Shaw IC, McCully S (2002)The potential impact of dietary en-docrine disrupters on the consumer. Int. J. Food Sci. Technol. 37, 471–476).

Food xenoestrogen	Source	Theoretical blood level (ng/l)
Natural		
Coumesterol	Beans, peas, spinach, soy	14
Genistein/genistin	Soy	143
Plasticisers		
Bisphenol-A	Canned food	0.1
Phthalates	Packaged food	0.003
Pesticides		
DDT, dicofol, endosulphan, dieldrin, β-HCH	All food	0.005

It is clear from this simple model that natural xenoestro-gens are more important than synthetic xenoestrogens as di-etary sources of estrogenic compounds. But, is there enough of them in men's diets to cause feminisation? If their total estrogenicity is greater than the normal circulating levels of 17β-estradiol they might. The normal level of 17β-estradiol in men is 20 ng/l, so they might well have a pharmacological effect because their theoretical estrogenicity calculated above is >150 ng/l.

What about the effects on women? Using the same philosophy we should compare dietary-derived xenoestrogen blood levels with normal female 17β-estradiol levels. This is very difficult because there is no normal level. Estrogen levels fluctuate sharply during the oestrus cycle – they can get as high as 5,000 ng/l. The xenoestrogen level contributed by food is a drop in this ocean of estrogens and is very unlikely indeed to have any pharmacological effect at all. So the xenoestrogen effect seems to be a problem restricted to men.

Barbara Thomson, one of my brightest PhD students, has done some nice work on New Zealanders' intakes of xenoestrogens. She showed clearly that the theoretical blood levels can be many times greater than actual xenoestrogen blood levels. Despite this, her calculations show that the total intake of xenoestrogens by New Zealand men could well have a pharmacological effect. The evidence is mounting!

Xenoestrogens and Breast Cancer

When a breast cell undergoes a cancerous change this is only the first step in the development of breast cancer. The cell must divide for the cancer to grow and spread. Breast cells are stimulated to grow by estrogens, in particular 17β-estradiol. Indeed this is one of the important natural effects of estrogens especially at puberty, after childbirth and during breast feeding. There is concern that dietary xenoestrogens might stimulate breast cancer cells to divide so facilitating the growth of the cancer.

This is highly unlikely to be the case in pre-menopausal women because they will have high and very variable natural circulating estrogen levels. However post-menopausal women have much lower estrogen levels – they effectively become biochemically more like men when their estrogen synthesis drops at menopause. It is therefore feasible that dietary xenoestrogens might help breast cancer cells to grow in post-menopausal women.

While on the subject of estrogens and breast cancer it is interesting to note that the most important medicine used to treat breast cancer is Tamoxifen, and that it is a 17β-estradiol analogue that blocks the estrogen receptor so preventing estrogen getting in and stimulating division of breast cancer cells.

Precocious Puberty in Girls

Precocious puberty is the early onset of puberty, as evidenced by the premature development of secondary sex characteristics (e.g. breasts) in girls. It has been suggested that dietary xenoestrogens are contributing to the steady increase in the incidence of this problem. This is most certainly a possibility because until the onset of puberty girls have estrogen levels much the same as boys. At puberty their ovaries begin to synthesise more estrogen to begin puberty. If dietary xenoestrogens are high enough it is possible that they could trigger puberty.

High Soy Consumers

People from Asia eat much more soy (e.g. as tofu – fermented soy bean curd) than westerners. Is sperm count declining quicker in these people? Do Asian girls enter puberty earlier? Is the incidence of breast cancer in Asian women greater? Very little work has come out of Asia on these issues, one thing is for certain that studies on blood levels show that people from that part of the world have higher concentrations of xenoestrogens than westerners. So the cause is present, but is there an effect?

Soy Consumption in the West

If you ask most westerners how much soy they eat they'll say, "not much". This is not necessarily true, because people tend to equate soy consumption with recognisable soy – like tofu, or soy milk. However soy is added to so many foods. Bread often contains some soy flour, sausages might contain soy meal as a binder, convenience sauces can be thickened with soy flour, etc, etc. I bought a loaf from my local supermarket this morning, the ingredients label reads:

> *Wheat flour, water, baker's yeast, salt, soy flour, vegetable oil, emulsifier (471, 472e, 481) acidity regulators (260, 262, 263). CONTAINS WHEAT AND SOY*

Most of us eat much more soy, and therefore soy xenoestrogens, than we think. This is evidenced by the enormous increase in soy acreage planted in the USA – in 1971 there were 43.5 million acres planted, by 1996 this had increased by 48% to 64.2 million acres. This increased production must reflect an increased demand. Some of this increased productivity will go to animal feed manufacture, and soy oil will be made from a proportion of the yield so not all of it will add end up on our plates as soy containing xenoestrogens.

Soy Oil

Soy oil is pressed from soy beans, and is increasingly found in our kitchens. Because the xenoestrogens are present in soy mainly as sugar conjugates they are very water soluble and therefore do not dissolve in the oil. For this reason soy oil is a poor source of dietary xenoestrogens.

Soy Milk and Infant Formula

Soy milk and infant formula are made by suspending ground up soy (soy solids) in water-based mixtures to look and taste like milk. There has been much controversy about the health effects of these products, particularly to infants and young children. We don't want to dose our little boys with xenoestrogens during their growth period for fear of affecting their sexual development.

There has been some excellent work conducted in the USA, UK and New Zealand on infant formula which all came to the same conclusion. Levels of xenoestrogens in infant formula are high, infants absorb them, and infants on soy formula have elevated blood xenoestrogen levels. A typical soy formula-fed infants xenoestrogen intake is similar to that of an adult from Japan (where soy is an important part of the diet), and higher than the level that is known to have an effect upon the hormone balance of women. This all sounds very worrying, and points to banning soy-based infant formulae … but let's explore a little further before we jump to conclusions!

Infants don't have as many active ERs as adults, therefore they might not have a great pharmacological response to estrogens. In addition, they are used to estrogen exposure from mothers' milk and via the placenta during gestation. Perhaps they have an inborn mechanism to resist the feminising effects. Who knows?

In my opinion we should operate the precautionary principle. Don't feed soy-based infant formula unless there are medical reasons to preclude mother's or cows' milk – i.e. the risk of feeding mother's milk is greater than the benefits. Mother's milk is best, it contains antibodies to immunise the baby against diseases to which the mum has been exposed, and it is a tailor-made nutrient source for human babies that does not need mixing, warming, or sterilizing!

We should not substitute soy impostors for their "real" counterparts (e.g. soy milk for cows' milk) unless we have some medical reason (e.g. allergy to cows' milk) to do so. The strict vegetarians and vegans will take me to task for saying this, but I challenge them not to consume things that look like animal products if they are opposed to eating animals (I only eat humanely produced (i.e. free range) meat so I am not as biased as you might think!).

Beneficial Effects of Xenoestrogens

This book focuses on an holistic view of food safety. I have tried not just address the risks of food, but have also put forward the benefits. Having both halves of the risk-benefit equation helps us to decide whether to accept the risk or not.

Xenoestrogens in food are perhaps the best illustration of good and bad in the same molecules. On the bad side, it is likely that xenoestrogens in food are feminising males (e.g. reducing sperm count). But there is a very important good side that must be considered before yelling, "ban xenoestrogens!"

Xenoestrogens and Osteoporosis

At menopause women reduce their synthesis of estrogens which results in loss of the important effects of these hormones. For example, estrogens are responsible for maintaining bone density – they facilitate the laying down of calcium in bone so keeping bones strong. When the influence of estrogens subsides at menopause, bone calcium might also diminish, so bones become less hard and are more easily broken. This is the basis of osteoporosis. The time taken for the decreased laying down of calcium to have an effect is long, so most cases of osteoporosis don't come to light until the patient is in her late 60 s or older.

A small dose of estrogens will reduce the reduction in bone calcium, so it is arguable that xenoestrogens in food will prevent (or ameliorate) the onset of osteoporosis. There is good scientific reason for thinking this, but very little hard scientific evidence. Despite this, some food manufacturers have marketed soy-containing products (e.g. breads) as being beneficial to women of "a certain age". They certainly won't do any harm, so why not eat them in the hope that they will be beneficial? There is just a tinge of worry about this philosophy. Hormone replacement therapy (HRT) is used to prevent osteoporosis (and other disorders) in postmenopausal women – HRT is the administration of tablets containing 17β-estradiol. However recent studies have shown an increased incidence of breast cancer in women receiving HRT; this is hardly surprising since some breast cancer cells are stimulated to divide by estrogens (i.e. they are ER positive cancers – see above). It is just possible that the xenoestrogens in food might result in an increased incidence of breast cancer in postmenopausal women. You will remember the discussion above where I dismissed the adverse effects of xenoestrogen in women because of their naturally huge levels of estrogens...but postmenopausal women don't have huge estrogen levels, in fact in this respect they are hormonally more like men – so it is possible that cancer incidence might be an adverse effect of "functional food HRT".

The potential benefit of xenoestrogens to women with osteoporosis is just one example of a possible benefit of xenoestrogens in food. As with most functional foods there is very little, if any, scientific evidence for their effect. The hype is around clever marketing. This does not mean that they are not beneficial, but it does mean that there is no hard evidence for the benefit half of the risk-benefit equation.

9
Genetically Modified Food

9   Genetically Modified Food

Since the Mad Cow saga subsided, genetically modified food (GMF) has taken its place as the *cause celebre* of food activists. The question is, are they justified in their vehement opposition to this amazing new technology? Should we join them, or ignore them? This chapter will introduce the issues and attempt to give an unbiased viewpoint on the food safety implications of GMF.

What is Genetically Modified Food?

Genetically modified, or engineered foods (they are the same – the opponents tend to use "engineered", the supporters prefer "modified") are simply foods that have been produced using alien genes inserted into crops or animals to introduce a useful production or flavour characteristic. For example insertion of a gene for herbicide resistance into corn means that the farmer can spray his crop with the herbicide to kill the weeds infesting the crop, but have no effect on the crop – more of this later. This makes the farmer's job much easier and will increase his yields (and income) by reducing weed competition. The problem, of course, is that when we eat the corn we also eat the inserted gene. Our question is, what will this gene do to us? Will it make us herbicide resistant? Will it harm us in some, as yet, unknown way? How dare they feed us genes that we don't normally eat!

There are several important GMFs that are grown around the world, and there are many more in the wings, and even more possibilities. We will look at these in detail later in this Chapter, but first a brief lesson on molecular genetics is necessary in order to understand what genetic modification is.

Molecular Genetics ... an Idiots Guide (Where the Idiot is the Guide!)

The cell nucleus is where the genetic material is, and where the blueprint of the cell is maintained. The genetic material is deoxyribose nucleic acid (DNA); it is composed of a phosphorylated sugar (deoxyribose) backbone with one of four DNA bases attached to each sugar unit. It is the sequence of these bases that determines the genetic code. The bases are thimidine (T), cytosine (C), guanine (G) and adenosine (A). In groups of three (codons) they code for amino acids which in turn are linked together to form proteins. For example, GCA codes for the amino acid arginine. Proteins (e.g. enzymes) control cellular activity and are important structurally; so changing the proteins can have a profound effect on cell function. Since proteins are built up from thousands of amino acids, it takes a long string of codons to code for a whole protein. This long string of codons is a gene (Fig. 9-1).

To translate the code in a gene into a protein there is an almost unbelievable sequence of biochemical events. It is at this point that I marvel at biology – this process developed randomly, with no help!

Translating the Genetic Code into Proteins

The DNA with its codons is coiled up and inactive in the nucleus. It remains in this form until the cell divides or protein synthesis is switched on. During this dormant period complex biochemical processes are in operation to repair damaged sections of the DNA. This is necessary because the DNA receives a continuous onslaught of damaging chemicals (e.g. carcinogens – some might come from food) and radiation (e.g. high energy

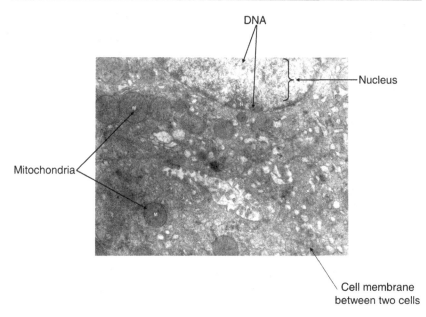

Fig. 9-1. An electron micrograph of a mammalian cell (from a rat) showing DNA in the nucleus – from Shaw, IC PhD Thesis, University of Birmingham, UK (1981)

γ-radiation from outer space) that knock out bases and so change the codons by shifting the reading frame – this is a mutation.

A frame-shift mutation:

DNA sequence: -GCA-CCT-ATG-TGA-AAA-CGG-CAA-AGA-

Protein sequence: -Arginine-glycine-tyrosine-threo-nine-phenylalanine-alanine-valine-serine-

Remove one base by mutation, the sequences becomes:

Mutated DNA sequence: -GCA-CCT-AGT-GAA-AAC-GGC-AAA-GA

Mutant protein sequence: Arginine-glycine-serine-valine-asparagine-glycine-lysine-

The mutated gene produces a protein that is totally different, and will have different (if any) activity to the original protein.

To translate a gene sequence into an amino acid sequence (protein), ribosenucleic acid (RNA) is made from DNA in the nucleus. RNA differs from DNA by having ribose instead of deoxyribose as its sugar backbone, and the base thymidine (T) is replaced by uracil (U) – so in the genetic code just replace the Ts with Us. The RNA passes out of the nucleus through the nuclear pores and moves towards structures in the cell called ribosomes – this is where translation of the genetic code to synthesise proteins occurs. This is not the place for a detailed discussion on protein synthesis, but believe me it is truly fantastic.

From the point of view of understanding GMF we need to appreciate the concept of the gene, because it is genes that are transplanted from one species to another to create the GMF crops. The genes can in theory be taken from anywhere. For example, the gene for herbicide resistance discussed above was taken (excised) from a soil bacterium and put into (implanted) a corn cell. The gene codes for enzymes that degrade the herbicide (glyphosate – Round-up) to which the corn is resistant. Clever stuff!

Moving Genes Around

Surprisingly it is relatively easy for molecular biologists to shift genes from one species to another. They can do it in several ways. For example viruses that excise the gene and incorporate it into their genetic material can be used. The virus containing the desired excised gene from species 1 (e.g. the soil bacterium in the glyphosate resistant corn example) is then allowed to infect species 2 (e.g. corn) and when it incorporates its nucleic acid into its host cell's genetic material (genome), as part of the infection process (see Chapter 4) it transfers the excised (e.g. glyphosate resistance) gene at the same time. The infected cell now takes on the properties of the excised gene (e.g. glyphosate resistance). When the cell divides, its "offspring" will also contain the gene and express the gene's properties.

Cloning

This is a way of reproducing an organism that does not involve sexual reproduction and maintains the same genetic code. For example, a plant normally reproduces by making ovules that are fertilized by pollen to become seeds, the seeds germinate to form a new plant that has some of the characteristics of the plant from which the ovule was derived and some characteristics of the pollen plant – cross pollination is important in maintaining genetic diversity as a means of evolution. However if you take a cutting (interestingly the word *clone* is derived from the Greek *Klon* meaning twig) the resulting plant will be identical to its parent (you would be shocked if a cutting from your scarlet geranium bore white flowers!). Cloning takes this concept one stage further. Individual cells are taken and grown up in culture media to produce adult plants (many of the house plants that we buy today are produced in this way). Millions of identical adults can be produced. They can be allowed to mature and form seeds (by self pollination, or cross pollination with pollen from other plants from the same clone) that will germinate to produce more identical plants. This is the way that molecular biologists and plant breeders produce GMF crops.

All of these principles can also apply to animals (remember Dolly the cloned sheep), but as yet there are no important GMF animals – although a GMF salmon is just around the corner.

Genetic Modification is Just Speeding up the Plant Breeding Process ... But is It?

Some protagonists of GMF production will present eloquent arguments about simply speeding up the traditional plant or animal breeding process. They say that breeders have since time in memoriam selected-in desirable genes (e.g. the genes for redness in tomatoes) to produce better crops that are more appealing to the consumer. This is, of course, absolutely true. However these selection processes while speeding up the evolutionary process never introduced genes that were alien to the plant or animal – this is simply not possible. All of the currently (2004) available GMF crops contain genes that have been transferred from one species to another (e.g. glyphosate resistant corn contains the glyphosate resistance gene from a soil bacterium). So this argument is not a good way of supporting GMF. Don't listen to anyone who tries to present it to you.

There is no reason why breeders could not assist, using molecular technology, the selection process for an endogenous gene. If they did it would be very difficult indeed for anyone not to accept the product even though it would be genetically modified. But as yet this has not been used to engineer a potentially important crop or animal for food production.

Genetically Modified Crops

Ethics

The mega-agrochemicals companies based in the USA have invested multiples of $billions in the research that has led to the production of genetically modified crops, and then similar enormous amounts of money in developing the research output into commercially viable crops. They will not accept argument against them without a very significant battle. They are immensely powerful and have political support in some countries. On the other hand they are based on science, and they employ some of the world's top molecular scientists who are ethical and accept good scientific debate. They are not trying to produce a monster that will destroy the world as we know it. But there must be tension between the massive investment and the need for payback soon which could cloud the ethical views of the leaders of these companies. No one outside the companies can ever know what debate goes on within their walls about ethics, and potential effects of GM technology, but one hopes that it is given time and space to consider the issues properly. I must say that when I have interacted with scientists from the GMF development and production companies I have been impressed by their integrity and desire to discuss the issues openly.

Crops in the Market Place

There are 6 genetically modified crops currently (2004) available showing how they might get into our food.

The genetically modified crops – information from: GM foods and the consumer, Australia New Zealand Food Authority, Canberra, Australia (2000)

Crop	Genetic trait	Potential food use
Soybean	Herbicide (glyphosate) tolerance High oleic acid production	Soy flour, soy milk, tofu, soy oil
Canola (oil seed rape)	Herbicide (glyphosate) tolerance	Canola oil
Corn	Insect protection (BT toxin) Herbicide tolerance (glyphosate) Both insect protection and herbicide tolerance in the same plant	Corn kernels, corn flour, corn oil Syrups made from corn starch
Potato	Insect protection (BT toxin) Insect protection + virus protection	Potatoes, potato products
Sugar beet	Herbicide tolerance (glyphosate)	Sugar
Cotton	Insect protection (BT toxin) Herbicide tolerance (glyphosate)	Cotton seed oil – used in some blended cooking oils

Herbicide Resistant Crops

As discussed at the beginning of this Chapter introduction of an herbicide resistance gene into a crop makes controlling weeds in the field very easy; the weeds die but the crop is unaffected. This is most farmers' dream.

There are herbicide resistant GM versions of soy, canola (oil seed rape), corn, sugar beet and cotton; they all have the glyphosate resistance gene from the soil bacterium *Agrobacterium tumefaciens*. The glyphosate resistance gene codes for an important plant enzyme 5-enolpyruvylshikimate-3-phosphate synthetase (EPSPS) which catalyses the synthesis of aromatic amino acids which are important in the synthesis of proteins. Animals don't have this enzyme because they get their aromatic amino acids from plants in their diet. Glyphosate works by inhibiting EPSPS so preventing the plant producing important amino acids and thus killing it. Monsanto found that *A. tumifaciens* EPSPS is resistant to the inhibitory effects of glyphosate (it has a different amino acid sequence probably due to a mutated gene), they excised the *A. tumifaciens* EPSPS gene and transferred it to other plant (e.g. soy) cells. When the GM cell was cloned the plant it produced synthesized both the glyphosate sensitive EPSPS translated from its own gene plus the glyphosate resistant EPSPS from the *A. tumifaciens* gene. The result was a plant that could synthesise aromatic amino acids in the presence of glyphosate; this meant that it was glyphosate resistant (or tolerant), Fig. 9-2.

Once the process had been worked out it was easy to transfer the *A. tumifaciens* EPSPS gene to any plant as a means of introducing glyphosate resistance. We now have cotton, soy, sugar beet, and canola ... and there are probably more in development as I write.

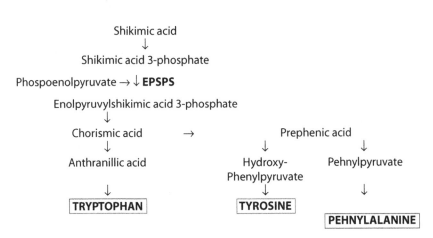

Fig. 9-2. Synthesis of the aromatic amino acids, tryptophan, tyrosine and phenylalanine in plants by the shikimate pathway showing the importance of the glyphosate inhibited enzyme 5-enolpyruvylshikimate-3-phosphate synthetase (EPSPS)

Insect Protection

Both the use of pesticides and their resultant food residues (see Chapter 7) are controversial, therefore any means by which we can reduce pesticide use must be considered seriously. The production of insect resistant crops is one way of achieving farming without pesticides. The crop itself kills insect pests!

Just in the same way that the glyphosate resistant crops were produced the discovery that a toxin produced by *Bacillus thuringiensis* (BT toxin) is lethal to insects at very low dose was seized upon, the gene identified, excised, and transferred to plants. The GM plants synthesized (expressed) the toxin and were toxic to insects – brilliant! The toxin works by paralyzing the insect's digestive tract so preventing it from eating and starving it to death. It works best during the insects larval stages because larvae usually eat more than adults (because they are growing). BT toxin has been very successfully used for quite a long time as a spray; it has the added benefit of being virtually non-toxic to mammals and birds.

BT-toxin genes have been incorporated into corn, potatoes, and cotton crops and all are now grown commercially.

Virus Protection

There are a number of viruses that are a significant threat to the potato industry, for example the potato leaf roll virus (PLRV). Resistance to the virus can be conferred upon potatoes by a genetic modification involving insertion of the gene for the viral coat protein (excised from the viral genome) into the potato. This is a modification specific to the potato.

Do All GMFs Contain Alien Genes?

Many people are worried about eating GMF because the food might contain the alien genes inserted as part of the genetic modification process. There is no evidence that these genes will do any harm to consumers. Indeed there is a steadily increasing battery of evidence from animal experiments that they have no effect whatsoever. A few experiments have shown effects in mammalian consumers, and these have received a disproportionate amount of press. For example, studies by Dr Arpad Pusztai at the Rowett Institute in Scotland suggested that rats fed on GM potatoes had significant differences in organ weight and depression of lymphocyte (white blood cell) responsiveness compared to controls. There was an enormous dispute about these data which led to Dr Pusztai's dismissal from the prestigious Rowett Institute. A group of eminent scientists reviewed the studies and concluded that they were preliminary, but were scientifically robust – preliminary usually means that more experiments are needed to prove the point. There have not been more experiments.

GMFs can be divided into two distinct classes:

1. Foods containing alien genetic material (e.g. corn starch from GM corn)
2. Foods not containing any alien genetic material (e.g. canola oil from GM oil seed rape)

Foods in class 2 cannot result in ingestion of alien genetic material because they are derived by processes that separate non-nucleic acid containing products from the GM plant. For example, oils are pressed from plants (e.g. canola), the oil is filtered and contains no plant material. Furthermore nucleic acids are not oil soluble and so cannot contaminate the product.

Other products of GM crops do contain nucleic acids (class 1 above). Eating the whole plant (e.g. potatoes) will obviously result in ingestion of the alien gene. Corn starch and soy flour will also include the alien nucleic acid because they are made by grinding the seeds from the GM crop, and the seeds are composed of cells with nucleic acid-containing nuclei.

Non-Nucleic Acid Differences Between GMF and Conventional Food

Insertion of a new gene into an organism might have all sorts of effects on cellular metabolism. It might result in changes in the emphasis on certain biochemical pathways, it might cause the switching on of inactive metabolic pathways, it might do almost anything; this is uncharted ground. Dr Pusztai's work in Scotland showed differences between the chemical make up of control and GM potatoes. Whether these data are reliable is uncertain, and whether the biochemical changes are of concern to the consumer is a moot point. The regulatory authorities in most countries attempt to assess nutritional and chemical similarity between conventional and GM crops before approving the latter for consumption. There are often differences – I have been involved in assessing this for Food Standards Australia New Zealand (FSANZ), but what does a small change in the level of a particular sugar mean when the natural variability between conventional crops grown in different places is large? I suspect that these differences are irrelevant. It's a bit like trying to regulate a cup of tea based on its sugar content... some people like one spoon, others like two; this does not mean that the people who take two will suffer deleterious effects. This is an area of considerable debate, we will probably never resolve it to everyone's satisfaction.

GMF Legislation Around the World

Different countries have different legislative views on GM crops. Some allow them to be imported as GMF, but don't allow the GM crops to be grown, others don't allow import or growth, others allow both import and growth. However, irrespective of whether your country allows import and/or growth or neither, you will almost certainly have consumed some alien genes in your food because it is very difficult to segregate GMF products and non-GMF products – they look, feel, and taste just the same. Many countries have monitoring programmes in support of GM labelling legislation and so have an assessment of food chain contamination with GMF in their country.

GMF in Food in New Zealand

New Zealand is a good example to illustrate the issues surrounding GMF in food. Until very recently GM crops were not allowed to be grown in the country (supervised, highly regulated growth as part of research and development work is now allowed – this was an extremely controversial move, and in my opinion wrong ... let's keep somewhere free of gene insults), and GMF was not permitted to be imported or incorporated into food products to be sold within the country. New Zealand recently introduced labelling regulations that require all products containing GMF to be labelled accordingly. If a product is labelled GMF-free it must be made from known GMF-free components. Unlabelled foods are not allowed to have GMF components greater that 1% of the product. A monitoring programme was introduced to ensure that the regulations were being followed. Of 103 samples analysed in the early 2000s, 16 (15.5%) were found to contain genetic modifications. The positives included 6 samples of corn chips, a vegetarian sausage, preserved soy in rice sauce, and a soy-based infant formula. Whether this frequency of positives is acceptable is up to you to decide. Surprisingly the New Zealand press hardly mentioned the findings.

Is it Safe to Eat GMF?

We eat genes in our food every day; genes from the vegetables, fruit, and meat that make up our diet. Eating genes from an orange does not mean that we will turn into a citrus tree, so there is no reason why eating any other gene will lead to any effects different to those of the other genes alien to our bodies that we consume daily.

The biochemical differences between GMF and its conventional counterparts are small and generally within the variability that we might expect between conventional crops grown in different places and at different times of the year.

I am not in the slightest bit worried about eating GMF. I do not believe that it will have any adverse effects on me.

Environmental Impact of GM Crops

Transferring genes from bacteria and viruses into plants and then releasing the plants into the environment will undoubtedly result in the transferred genes being carried away from the field in which the crop is growing in pollen. This will only be important if the pollen can fertilise another plant so resulting in the alien genes being incorporated and possibly expressed in another plant's genome. This could only happen if the GM crop's pollen alighted onto a closely related plant, and if the pollination resulted in a fertile seed that germinated. One of the strategies to prevent this happening has been to grow crops in countries where there are no plants closely related to the crop species – pollination will only occur between the same or very closely related species. However as GM crops become more common this ideal is being eroded. For example GM oilseed rape is now growing in countries where wild mustard grows. Mustard and rape are closely related and therefore cross pollination might occur. Similarly GM corn was initially grown in the USA where wild maizes do not grow. GM corn is now being grown in parts of the world where plants closely related to maize are native. If cross pollination to produce viable offspring does occur, the glyphosate resistance gene might contaminate the wild gene pool and result in weeds becoming glyphosate resistant too.

More worrying is the transfer of insect protection genes (e.g. BT toxin) to wild plants. If these plants are insect food plants the effect could be devastating. The problem is that we know too little about these possibilities to be certain. Surely we should not release these genes in a mobile form into the environment until we understand their behaviour and spread. Alas GM crops are big business and there is a very strong commercial lobby pushing for their release and trade.

There is significant concern amongst scientists about the possibility of horizontal gene transfer. This is the transfer of a gene from a GM crop plant to an intermediate species (e.g. a soil bacterium), followed by its transfer to another species. Very little is known about this except that it is possible in a culture flask. The effects of horizontal gene transfer could be serious because it does not rely upon similar species growing close together to result in gene transfer and expression in species unrelated to the original GM crop.

In my opinion the environmental argument against GM crops is strong. Not because we know that genes will be spread around the environment resulting in devastation, but because we simply do not know what might happen. Surely we should understand more before irreversibly releasing these potentially mobile genes into our environment.

10
Is it Safe to Eat?

10 Is it Safe to Eat?

And finally the big question, **is it safe to eat?** One thing is certain, if you don't eat you'll die. But that is a little extreme because you can select your food to minimize the risk. So the answer to the question is *yes*, but you can make it safer by eating properly and wisely. Properly means a wide varied diet in order to honour Paracelsus and minimize your exposure to hazards so minimizing the risks. Don't eat the same thing over and over again during the same week. Recent worries about dioxins in farmed salmon illustrate this well. These environmental pollutants have found their way into salmon feed and are contaminating salmon meat at potentially unacceptably high levels, they cause cancer. The word cancer alone frightens most people off, so people have stopped buying salmon. They have not done a proper risk-benefit assessment! Too much salmon might harm you, but the occasional meal will have no deleterious effect whatsoever. In fact salmon contains a myriad beneficial fatty acids (e.g. omega fatty acids) that are important components of our diet, a lack of these is harmful. So be sensible, eat salmon, but only every other week. This, of course presents the salmon producers with a difficult marketing problem – how many marketers are experienced at telling their customers not to buy too much of something? But, how many marketers are food risk experts? Listen to the scientists not the marketers!

Nothing is safe, so to help us to decide whether it is safe to eat we need to set the risks of eating food in the context of the risks of living. I have discussed this already in Chapter 2 – you have far more chance of dying in a car accident than from food.

The benefits of food are great, it provides sustenance, essential micronutrients (e.g. vitamins), and sheer pleasure – just think of life without the joys of eating. We could make food much safer by sterilizing everything, reducing food to tablets of nutrients, calories, and micronutrients- all you need to survive, but not all you need to live – enjoyment is a key facet of life. We must accept the relatively low risks associated with eating in order to live, not just survive. But, we must also use our expertise and technologies to minimize the risks, but not reduce the pleasure. It was interesting in the UK when the government banned beef on the bone (e.g. T-bone steak) during the Mad Cow saga. They did this because of the miniscule additional risk posed by eating these cuts because nerves (where the BSE prion is) follow bones and therefore bone-in cuts are more likely to harbour prions. There was an outcry from consumers – we will eat what we like! We will decide whether or not to accept the risk! Don't tell us what to eat! This is not a nanny state! This is excellent evidence of assessment and acceptance of risk in light of enjoyment benefit. In other words the consumer is sensible and able to make their mind up about what is too risky to eat providing they are given the evidence.

The "Nanny State" Approach to Protecting the Consumer

I mentioned "the nanny state" above. It is important for governments to legislate to protect us from unacceptable risks, but they must educate us to assess low risks and make sensible decisions on these for ourselves – not everyone will make the same decision, because not everyone has the same acceptance of acceptable risk. The problem, of course, is that it is difficult to define the line between acceptable and unacceptable. I think that the UK government found it when they banned beef on the bone. But they listened and learned, and responded by rescinding the ban. Good on 'em!

UK Plans to Ban Beef on the Bone
Financial Times, December 4, 1997

Beef on Bone Ban to Stay
Guardian, February 4, 1999

Beef on the Bone Ban Lifted
MAFF, UK Press Release November 30, 1999

There are many more examples of the fine line between acceptable and unacceptable risk. They are all different, but there is one constant, there is little or no agreement on where the line should be – the best way forward is to educate and inform. Keep the consumer abreast of current knowledge, and let them make up their own minds, just like they decide whether or not to cross the road – we haven't banned crossing the road yet, and this activity is far more risky than anything that food has to offer.

The Benefits of Food

Before we can properly answer the question, *is it safe to eat?* we must not only dismiss the risks as being insignificant in the context of the risks of daily life, but must also set them in the context of food's benefits. I embarked upon this approach above when I entered the philosophical world of enjoyment as a benefit of food. It most certainly is, and therefore we must put this on the benefit side of the RISK vs BENEFIT equation, but it is difficult to quantify, and it is equally difficult to decide if enjoyment is sufficient to make risk acceptable. I think that it is, but some might not. So we need to search for the benefits of food and add them to our decision making process.

The Anti-Cancer Properties of Food

It is well known that certain chemicals reduce the risk of getting cancer, this has been demonstrated to scientists' satisfaction in cell culture and animal experiments. Even human epidemiological studies have shown links between consumption of certain foods and low cancer incidence. Examples of anti-cancer chemicals are the anti-oxidants such as vitamin C.

To understand this we need a quick lesson in carcinogenesis. Cancer is the uncontrollable growth of cells that results in their infiltrating normal tissues, robbing nutrients and grossly affecting the body's equilibrium eventually overgrowing the system and resulting in death. This is an all too familiar scenario. Everyone knows someone who has died of cancer, and most of us are amazed at the short time between diagnosis and death. Cancer cells grow fast. But what transforms a normal cell into a cancer cell? The answer to this question deserves an entire book. In the simplest terms the cell's DNA is altered in such a way as to remove any control on cell division so resulting in uncontrolled cell proliferation (some transformations do not involve DNA [i.e. non-genotoxic], but they are the exception – there is not space to go into this level of detail in this very cursory overview). This change is often caused by a chemical interacting with DNA to change its function. Many of these chemicals are activated by the body's own metabolic systems to highly reactive forms (species) that react quickly and irreversibly with DNA. Many of these reactive species are strongly oxidizing (e.g. free radicals). It is for this reason that anti-oxidants protect cells against some chemical carcinogens.

Many foods contain antioxidants (e.g. vitamin C in oranges) and so inhibit oxidizing chemical-mediated carcinogenesis. If you put vitamin C into the Ames test with powerful oxidation-mediated carcinogens, the test is negative – the Ames test is a test for mutagens (i.e. chemicals that alter DNA – most carcinogens are mutagens) invented by Professor Bruce

Ames from the USA. This is just the tip of the food anti-cancer benefit iceberg. Many more protective chemicals in food are being discovered as I write. Broccoli contains isothiocyanates which are potent cancer protectants; watercress contains similar chemicals – in fact its sharp taste is due to them; tomatoes contain anti-cancer lycopenes; fish contains vitamin E which is a powerful anti-oxidant. The list is long, and is steadily increasing.

I discussed functional foods in Chapter 6, but it deserves re-airing in the context of food benefits. The anti-cancer foods are functional foods, and this represents a significant benefit to the consumer. Food is inextricably linked with health. On the negative side it causes disease (e.g. heart disease due to its lipid content), but on the positive side it prevents, or some might say cures, disease. This is a huge benefit by anyone's standards.

Food can also prevent cancer by other mechanisms. For example, consumption of a fibre-rich diet is associated with a lower incidence of colon (lower bowel) cancer. The reason for this is not yet fully understood, but could involve the fibre promoting faster movement of faeces through the lower gut so minimizing exposure of colon cells to carcinogens contained in the faeces – this is just an idea, but gaining traction amongst scientists. So fibre's positive effects are a benefit of food.

Some races have different incidences of cancers; they also have different diets. It is tempting to associate their diets with these differences. My research group has studied New Zealand Maori's lower susceptibility to colon cancer compared to European New Zealanders (Pakeha). Many Maori still eat traditional foods such as watercress and puha (sow thistle – *Sonchus oleraceus*) – although puha is hardly traditional. It was taken to New Zealand as a contaminant of grass seed by the British settlers. We think that these "traditional" foods contain cancer protectant chemicals that reduce the rate of colon cancer despite Maori consuming more red meat than Pakeha which is associated with colon cancer. The following table shows colon cancer rates in New Zealanders compared with their consumption of traditional Maori foods and red meat.

Colon cancer rates in New Zealanders compared with their consumption of traditional Maori foods and red meat (data from Thomson, B & Shaw I (2002) A comparison of risk and protective factors for colorectal cancer in the diet of New Zealand Maori and non-Maori, Asian Pacific Journal of Cancer Prevention 3, 319–324)

	Maori	Non-Maori
Colon cancer rate (1998) (per 100,000 population)	22	44
Red meat consumption (g/day)		
Males	98	72
Females	65	42
% of population consuming traditional Maori foods		
Puha	14	0
Water cress	15	1

There are also numerous examples of race-related cancer trends linked with cancer causing foods. For example, the incidence of oesophageal (gullet) and stomach cancers are greater in Japan than anywhere else in the world. This has been linked to the Japanese's penchant for smoked food – smoked food contains the potent carcinogen benzo[a]pyrene. The following table shows the incidences.

Incidence of oesophageal cancer in Japan and USA (data for JAPAN (2001) from http://www.ncc.go.jp; USA (1996–2000) from http://www.seer.cancer.gov)

	Cancer incidence (per 100,00 population)	
	Oesophagus	Stomach
Japan		
Male	17.1	68.5
Female	3.0	34.4
USA		
Male	8.1	12.2
Female	2.1	5.7

It is interesting that the incidence of stomach cancer in Japanese Americans (males and females) is 28/100,000 which is lower than the incidence for Japanese in Japan. This suggests that the change to a US diet has had an effect. This is a positive effect of a US diet – there are many more statistics that show that its high fat content results in an increased incidence of coronary heart disease. This illustrates the danger of separating effects of food on health. Food must be considered holistically – i.e. risk and benefit together.

Other Health Benefits of Food

One of the great health concerns of our times is coronary heart disease. There is no doubt whatsoever that food influences this due to its fat content. High fat diets are associated with coronary heart disease. Cholesterol is a key determinant of coronary heart disease. Some foods however protect against heart disease. For example, people who eat more fibre in their diets tend to have a lower incidence of heart disease. There are a multitude of scientific explanations for this. It might simply be that people who consume more fibre are eating a greater proportion of fruit and vegetables in their diets and so not eating as much animal fat...or it might be that fibre protects in its own right. There is some evidence that specific components of high fibre foods prevent the absorption of high heart disease risk dietary components – such as cholesterol. Oats, for example, contain elements that do just this. Calcium has a similar effect. Therefore the cholesterol in milk might not be as bad as we thought because the high calcium content of milk reduces its bioavailabililty. Similarly a slice of your favourite cheese on a delicious Scottish oatcake might be safer because the oats reduce the absorption of cholesterol from the cheese. A bowl of steaming porridge before a hearty, fat-rich English breakfast might just reduce the health impact of the animal fats – and shows that Scots and English can collaborate beneficially! These examples illustrate very well the enormous complexity of dietary interactions.

Concern leading to government advice can have an enormous effect on a country's diet. In the UK in the 1980s there was concern about vitamin A in pregnant women's diets. Data from rat studies had shown that vitamin A can cause birth defects. Calculations showed that high liver consuming pregnant women might well consume enough vitamin A to be detrimental to their child. The concern was announced. Many pregnant women stopped eating liver altogether – then concerns about vitamin B12 deficiency in pregnant women became very real because a major source of this important vitamin is liver. This is another illustration of risks and benefits in a holistic context. Over reaction is just as bad as no reaction at all. Paracelsus realized this 500 years ago, will we never learn?

Cut the Waffle and Decide! Is it Safe to Eat or Not?

The greatest risk of eating is getting run over on the way to buy your food, not from the food itself. There are myriad disease preventing and health promoting benefits of food. But there are also chemical and microbiological hazards lurking in every bite. And there is the sheer enjoyment of eating. Putting all of these factors together I have absolutely no doubt that it *is* safe to eat providing you eat a broad, varied and well balanced diet. This will minimise your exposure to food hazards which means that the risk will be so low that you might just as well forget it and enjoy your tucker!

It is safe to eat – Enjoy!

Subject Index

Subject Index